T0192446

Technical Writing

A Practical Guide for
Engineers, Scientists, and
Nontechnical Professionals
Second Edition

What Every Engineer Should Know

Series Editor
Phillip A. Laplante
Pennsylvania State University

For more information about this series, please visit: www.crcpress.com

Technical Writing

A Practical Guide for Engineers, Scientists, and Nontechnical Professionals
Second Edition

Phillip A. Laplante

CRC Press
Taylor & Francis Group
Boca Raton London New York

CRC Press is an imprint of the
Taylor & Francis Group, an **Informa** business

CRC Press
Taylor & Francis Group
6000 Broken Sound Parkway NW, Suite 300
Boca Raton, FL 33487-2742

Printed on acid-free paper

International Standard Book Number-13: 978-1-138-62810-6 (Paperback)
International Standard Book Number-13: 978-1-138-60696-8 (Hardback)

Library of Congress Cataloging-in-Publication Data

Names: Laplante, Phillip A., author.
Title: Technical writing : a practical guide for engineers, scientists, and
nontechnical professionals / Phillip A. Laplante.
Description: Second edition. | Boca Raton : Taylor & Francis, CRC Press,
2018. | Series: What every engineer should know | Includes
bibliographical references and index.
Identifiers: LCCN 2018008664| ISBN 9781138628106 (pbk. : alk. paper) | ISBN
9781138606968 (hardback : alk. paper) | ISBN 9780429467394 (ebook)
Subjects: LCSH: Technical writing.
Classification: LCC T11 .L264 2018 | DDC 808.06/65--dc23
LC record available at https://lccn.loc.gov/2018008664

Visit the Taylor & Francis Web site at
http://www.taylorandfrancis.com

and the CRC Press Web site at
http://www.crcpress.com

This book is dedicated to Dr. Divyendu Sinha (1961–2010), a world-class scientist, writer, collaborator, and, most importantly, friend.

Contents

What Every Engineer Should Know: Series Statement

What every engineer should know amounts to a bewildering array of knowledge.

Regardless of the areas of expertise, engineering intersects with all the fields that constitute modern enterprises. The engineer discovers soon after graduation that the range of subjects covered in the engineering curriculum omits many of the most important problems encountered in the line of daily practice—problems concerning new technology, business, law, and related technical fields.

With this series of concise, easy-to-understand volumes, every engineer now has within reach a compact set of primers on important subjects such as patents, contracts, software, business communication, management science, and risk analysis, as well as more specific topics such as embedded systems design. These are books that require only a lay knowledge to understand properly, and no engineer can afford to remain uninformed about the fields involved.

Preface

You will expect a book about technical writing to have certain character-istics, but I doubt that "interesting" is one of them. Yet I hope to make this book technically correct, informative, and interesting—three features that do not readily coexist. Indeed, my goals are ambitious.

There are numerous editorial resources for writers, such as *Harbrace College Handbook* [Hodges et al. 1999], *The Elements of Style* [Strunk and White 2008], *On Writing Well* [Zinsser 2006], and *A Manual for Writers of Research Papers, Theses, and Dissertations* [Turabian et al. 2007]. Some standard texts on techni-cal writing also exist (e.g., Higham [1998]). I am neither a scholar of writing nor a professor of English or communications, so my point of view is unlike standard writing texts and reference guides. My intent is that this book complements the traditional writer's reference manuals and other books on technical writing. I also want this work to be compelling, even fun, to read, and so I have woven in a number of personal anecdotes, stories, and histori-cal vignettes.

My background is primarily in software and systems engineering, but I also have experience in electronic systems and hardware. I have spent vari-ous parts of my career in industry and academia and have participated in a wide range of activities in electrical and systems engineering. Many of the forthcoming examples are from my experiences in these domains.

I like to avoid long runs of text when I write. I am a "visual" person, so I try to intersperse at least one figure, table, graphic, bullet list, or equation on each page of writing. This approach helps make writing easier and read-ing more pleasant. I discuss the use of equations, graphics, and similar con-structs in Chapter 7.

I want to show you many examples of technical writing. It might seem immodest, but I draw upon my published works for examples. I decided to use my publications for several reasons: It would be easier to secure permis-sions, I could be critical without offending another writer, and I knew the back stories. These stories illustrate how the publication originated, various challenges and their solutions, and anecdotes, which I hope will keep your interest. I hope you will forgive me, then, if much of this book is semiautobio-graphical as I share the inside stories relating to the various writing samples.

New for the Second Edition

Since publication of the first edition, I've taught technical writing to many professionals and thousands of copies of the book have been sold. The book is used by faculty and hundreds of students at other universities. From these persons I have received many recommendations for improvement.

The second edition incorporates this feedback including new topics, additional examples, insights, tips and tricks, new vignettes, and more exercises. Appendices were also added to provide additional writing samples, templates for many kinds of documents and a writing checklist. The references and glossary have been updated and expanded. Finally, a perspective on writing for nontechnical persons has been incorporated.

Errors

I have tried to make this text accurate and correct, but I know from experience that no matter how hard I try to remove them, errors will remain. There is no more frustrating moment than to receive a newly printed book or paper fresh from the publisher, open it up, and immediately discover an obvious error.

Any errors of commission, omission, misattribution, etc. in this book are my responsibility, and I have an obligation to correct them. Therefore, please report any suspected defects to me at plaplante@psu.edu

References

Hodges, J. C., Horner, W. B., Webb, S. S., and Miller, R. K., *Harbrace College Handbook: With 1998 MLA Style Manual*, Harcourt Brace College Publishers, Ft. Worth, TX, 1999.

Strunk, W. and White, E. B., *The Elements of Style: 50th Anniversary Edition*, Pearson-Longman, New York, 2008.

Turabian, K. L., Booth, W. C., Colomb, G. G., and Williams, J. M., *A Manual for Writers of Research Papers, Theses, and Dissertations, Seventh Edition: Chicago Style for Students and Researchers (Chicago Guides to Writing, Editing, and Publishing)*, University of Chicago Press, Chicago, IL, 2007.

Zinsser, W., *On Writing Well, 30th Anniversary Edition: The Classic Guide to Writing Nonfiction*, Harper Books, New York, 2006.

Acknowledgments

Many writers have influenced me over the years. I read books from most genres, and my favorite authors include James Lee Burke, Bernard Cornwell, William Manchester, and Gore Vidal. For popular science writing, you can't beat Malcolm Gladwell or James Gleick. Many other authors have influenced my writing, but they are too numerous to mention. However, I would like to thank all of these named and unnamed authors for showing me how to write well.

During the writing of this manuscript, many people provided assistance to me in various ways. These individuals include:

My wife, Dr. Nancy Laplante, who is a terrific writer, for proofreading the manuscript, for discussing many ideas with me, and, of course, for her constant encouragement.

My son, Chris, for providing several examples, ideas, and technical support.

Acquisitions editor, Allison Shatkin, for her encouragement throughout the writing process.

My longtime friend and Taylor & Francis publisher, Nora Konopka. We have worked on many book projects together over the last 25 years.

Tom Costello, President and CEO of Upstreme, Inc., for sharing several ideas and writing samples.

Ernie Kirk, owner of Kirk's Premier Martial Arts, for allowing me to reprint excerpts from a *Krav Maga* instructor certification exam that he wrote.

Dr. Jim Goldman, for proofreading the first edition of this manuscript, making many excellent suggestions for substantive improvement, and providing material relating to cognitive authority in online publications, which was the subject of his doctoral dissertation.

Professor Paolo Montuschi, University of Torino, for contributing most of the discussion on e-readers in Chapter 9.

Dr. Jeffrey Nash, for contributing to the discussion on dynamic content in Chapter 7.

Dr. Colin Neill, for our many writing collaborations, excerpts of which appear in this book.

My former graduate student, Andrew Rackovan, for the extended example of Chapter 4.

Kellye McBride, for an outstanding job copyediting the manuscript.

Don Shafer of the Athens Group, for letting me share the evolution of our paper on the BP oil spill in Chapter 3.

The late Dr. Divyendu Sinha, for many things (see dedication).

Dr. Mitch Thornton of Southern Methodist University, for allowing me to share our press release in Chapter 5.

All of my other coauthors over the years, from whom I have learned much and with whom I have enjoyed great friendships.

I apologize if I have left anyone out.

1

The Nature of Technical Writing

1.1 Introduction

What is technical writing? I'm afraid that there is no universally accepted definition. Many authors discuss the difficulties in defining "technical writing" and then offer their own definition. For our purposes, it is easier to define technical writing by differentiating it from all other kinds of writing. There are two main differences between technical and nontechnical writing: precision and intent.

Precision is crucial in technical writing. When you express an idea in technical writing, it may be realized in some device or process. If the idea is wrong, the device or process will also be wrong. To quote my friend, physicist and software engineer par excellence, Dr. George Hacken, "syntax is destiny."

For example, imagine the consequences of an incorrect subscript in some chemical formulation, or a misplaced decimal point in a mathematical specification of some process for controlling a nuclear plant. Precision is particularly important in computer software. In 1962, a NASA Mariner 1 Venus satellite was lost, in part because of a misplaced hyphen in a data editing program [NASA 2017].

Precision in other kinds of writing is also important, of course. The title of Lynne Truss' book on punctuation, *Eats, Shoots & Leaves*, makes this point [Truss 2004]. The title refers to the dietary habits of a panda. However, if you add a comma after the word "eats," the title now could refer to a diner who refuses to pay his restaurant bill and shoots at the proprietor before fleeing the scene.[1] But the consequences of this kind of mistake are not nearly as potentially disastrous as in the specification, design, or code of some mission-critical system. Even in legal documentation, where imprecision can have deleterious consequences, there is not the same risk of loss of a system or life.

Another characteristic difference of technical writing is that there should be no intent to evoke an emotional response from the reader. The technical writer should simply try to convey information as concisely and correctly as possible. In poetry, prose, news reporting, and even business writing, it is necessary to convey information content or a story. But in poetry and prose, it is clear that an emotional response is also desirable. The situation is the same in news, where the reporter may be looking to scare, shock, or evoke sympathy or pity from the reader. Even in everyday business correspondence such as advertising, contracts, lawsuits, job applications, and so on, a visceral response or at least a call to action is desirable. This is not the case in technical writing.

A valid objective of technical writing may include persuasion of opinions, for example, convincing readers that a commonly held view about a topic is incorrect. Conveying neutral, but correct and concise, technical information often brings about this type of education in an unemotional and nonthreatening way.

Although they may not be truly "technical", encyclopedias, dictionaries, handbooks, directories, etc. fall under the category of technical writing. These items are truly "technical" in the sense that precision is needed.

You are likely to find equations or technical terms in technical writing—this situation is different from other kinds of writing. But equations neither define technical writing, nor necessarily do they define precision. Technical writing may exist entirely without any equations; for example, a guide may contain only step-by-step procedures for assembly, installation, use, or deconstruction of some product. Equations can also be imprecise or incorrect.

Finally, there are legal implications to technical writing. While any kind of writing can be libelous, an error in technical writing can have serious consequences. For example, writing quality in user manuals is known to have caused catastrophic software failures [Wong et al. 2017]. Other technical writing errors could lead to financial loss, damage to property, environmental catastrophe, injury, or death. Consider, for example, the potential consequences of the following: bad financial advice in an investment brochure, a wiring instruction error in a manual for an electric clothes dryer, an error in a hospital record for a seriously ill patient, an incorrect formulation recipe for mixing pesticides, or an error in the maintenance instructions for an aircraft.

1.2 Who Writes Technical Documentation?

I imagine that if you made a list of professionals who must write technically, you would include engineers, scientists, architects, physicians, lab technicians, and so forth. In the broadest sense, virtually any trade or profession can be considered to have a technical component, and its practitioners must prepare technical writings. Think about doctors, nurses, farmers, lawyers, and experts of all types. Every one of these persons will write in the jargon of their discipline—a kind of technical writing. From this point forward, when I say "technical professional," I mean a large and flexible collection of any profession or trade where technical writing can occur.

Everyone is a technical writer, at least occasionally. Product complaint letters, driving directions, or recipes written for friends are all kinds of technical writing. Disclosures to insurance companies, responses to legal inquiries, and incident reports at work should also be treated as technical writing—in these it is especially important to be very precise, include provable facts, and avoid expressing emotion. Whenever you endeavor to write something at work or elsewhere, pause to consider if that writing should be treated as technical.

1.3 Taxonomy of Technical Writing

For ease of discussion throughout the remainder of this book, I refer to the taxonomy described by Montgomery and Plung [1988], shown in Figure 1.1.

Pedagogically oriented technical writing focuses on teaching, for example, a calculus textbook or a book for the novice photographer. Technical writing of a theoretical orientation involves various kinds of theoretical and applied research. The broadest form of technical writing—professional orientation—serves the needs of various professionals. As has been mentioned, these

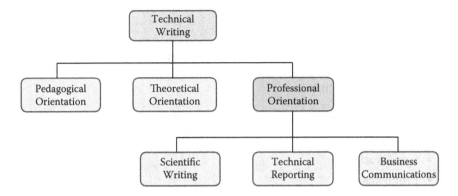

FIGURE 1.1
An illustrated taxonomy of technical writing. (Redrawn from Montgomery, T. and Plung, D., *Proc. of International Professional Communication Conference, 1988*, Seattle, Washington, October 5–7, 1988, pp. 141–146.)

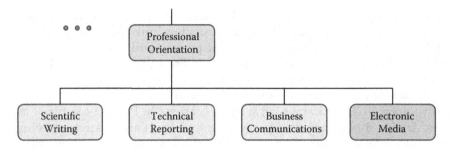

FIGURE 1.2
An updated version of Montgomery's taxonomy for technical writing of a professional ori-
entation. (Redrawn from Montgomery, T. and Plung, D., *Proc. of International Professional
Communication Conference, 1988,* Seattle, Washington, October 5–7, 1988, pp. 141–146.)

professionals may be in any discipline. Professional orientation is the class
of technical writing on which I will concentrate.

Although briefly mentioned by Montgomery and Plung, at the time their
paper was written, electronic media was very new. Since then, however, a
new form of written media and a unique style of writing have emerged.
I would like to expand Figure 1.1 to include these forms of professional
writing, adding a new category under "Professional Orientation" called
"Electronic Media" (see Figure 1.2).

Let's look at each of these areas under "Professional Orientation" in some
further detail. Examples of most of these various technical writing forms can
be found in later chapters in this book. I have organized these later chapters
to correspond with the major headings below.

1.4 Technical Reporting

Technical reports are documents that are prepared for supervisors, subordinates, peers, customers, clients, and various government agencies. Typical
technical reports include:

- Progress reports
- Feasibility studies
- Specifications
- Proposals
- Facilities descriptions
- Manuals
- Procedures

- Planning documents
- Environmental impact statements
- Safety analysis reports
- Bug reports

There are many other types of reports, of course, but all have a unity of purpose: to convey specific information in an archival way. By "archival" I mean that the document is intended to be stored and referenced for many years.

1.5 Business Communications

Business communications include a wide range of correspondence that must be written in the course of business activities. Typical business communications documents that you may read or write include:

- Résumés
- Cover letters
- Transmittal letters
- Customer relations writing
- Human resources communications
- Trip reports
- Administrative communications

Of course, there are other types of such correspondence that you may encounter in the workplace as a vendor, customer, client, consultant, or worker.

Here is an example, it's my short autobiography or "biosketch," which I use when submitting articles to journals and magazines. One of the challenges in writing a biosketch is keeping it short—they are often limited to 100 words or less (this example is 50 words):

> Phil Laplante is a professor of software and systems engineering at Penn State. His current work encompasses software testing, software security, requirements engineering, the internet of things, and software quality and management. He holds a PhD in computer science from Stevens Tech and is a licensed professional software engineer in Pennsylvania.

I discuss business communications, including résumé writing, in Chapter 5.

1.6 Scientific Writing

Scientific writing includes experimental research and associated documentation, as well as the scholarly publications that emerge from that work. Scientific writing also includes scholarly and experimental research in medicine.

Scientists and engineers can publish their work in a variety of venues, including:

- Books
- Journals
- Magazines
- Conferences
- Newsletters
- Websites and blogs

These kinds of technical publications vary widely in authority and in rapidity of publication. Authority refers to the reliability of the scientific content, which tends to be much higher if manuscripts submitted for publication are reviewed by technical peers and tends to be much lower if the writing is not peer reviewed. The speed with which material gets published is an important consideration of the timeliness of its content (Figure 1.3).

Peer review takes many months, sometimes a year or more for very complex technical material. Therefore, any research findings must be of an archival quality, meaning that the work must focus on long-term theoretical concepts that are not expected to change soon. Technical matter that will change rapidly must be published in venues that have a very short time from submission of the material to publication, or the technical writing will be stale before it appears.

Collectively, journals, magazines, and newsletters are referred to as "periodicals" because they are published at some regular rate, say from four to twelve times per year. Usually[2] these publications are digitally archived, that is, placed in an electronically accessible, Web-based library. Writers of technical material who wish to publish and users of technical material for reference must consider the trade-off of time to publication versus authority.

In Chapter 8, I describe in detail the type of writing used for these kinds of publications and give many examples. In the next few sections, I describe several types of technical writing, including books, journals, magazines, conference proceedings, newsletters, websites, and blogs. Please refer to Figure 1.3 as needed during these discussions.

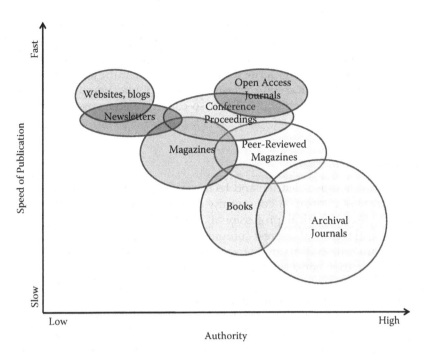

FIGURE 1.3
Speed of publication versus authority of content for a variety of technical writing types.

1.6.1 Books

Books are a way for scientists and engineers to get very wide and archival distribution for their ideas. But books take a long time to write, edit, and finally publish—often three years or more from the initial concept to the published book. If focusing on the professional reader, a fast publication time is critical, while for college textbooks, a longer period of development is acceptable. Books receive a certain amount of technical vetting through peer review.

1.6.2 Journals

Journals are the preferred venue for publication of important scientific ideas and technical breakthroughs. Scientists and engineers seek to publish in the best journals—those that are widely read and cited by other scientists and engineers. Journal articles usually focus more on theory and less on applications because they are intended to be relevant for a very long period of time. Theory changes more slowly than the applications for the theory.

When articles are submitted for consideration in a journal, they are reviewed by experts to determine if the work is worthy of publication.

This activity is often called "refereeing." It can sometimes take two years or more from submission to final publication of a paper in a scientific journal. So, although the article will exist in a digital library in perpetuity, timeliness is of concern for those who write for and publish scientific journals.

A new kind of journal with a faster time to publication and lower cost to readers has recently emerged. These so-called "open-access journals" are discussed in Chapter 8.

1.6.3 Magazines

There are numerous scientific and technical magazines catering to various communities of interest. A community of interest is a group with a shared focus, whether technical professional, political, recreational, religious, or other. Typical technical interest communities include green energy, robotics, personal aircraft, and history of science. Unlike technical journals, which tend to be rather conservative in appearance, technical magazines may resemble any magazine that you might find on a newsstand, complete with color graphics, advertisements, and editorials.

Magazine publications can be refereed, although this is not always so. The time from submission to publication for a magazine is usually much shorter than for a journal publication. Therefore, when scientists wish to get their ideas out quickly, but in a venue that has a certain amount of prestige, they may choose to publish in a widely read magazine.

Readers of magazines are generally looking to keep up with developments in their field without having to read the more theoretical papers that tend to appear in journals.

1.6.4 Conference Proceedings

Conferences are meetings where researchers present scientific findings, often in preliminary form. Experts are sometimes invited to these conferences to give presentations. In other cases, intended presenters must send a prospectus of what they intend to present (called an "abstract") or even a fully developed paper on which the presentation will be based. A vetting committee reviews the abstract or paper presentation to decide whether the authors should be invited to present their work to the conference.

Often the presentations at conferences are based on papers that are intended to be published in a transcript of the conference called "proceedings." The conferences are often peer reviewed, and the proceedings tend to be published within a few months of the conclusion of the conference, although there is a very wide range of practices and quality in conference proceedings publications.

1.6.5 Newsletters

Newsletters are informal publications produced by some community of interest, for example, a user group, a special interest group of some professional society, or an informal collection of practitioners. These publications have a fast turnaround, often only a few days, and are a way to expose an idea for rapid consideration and discussion. Newsletters, not usually peer reviewed, and do not have the prestige of journals, magazines, or conferences.

1.6.6 Websites and Blogs

Many technical disciplines, subdisciplines, and specialties have spawned one or more dedicated websites and columnist blogs. Websites and blogs can have nearly instantaneous publication so that they can be very timely in their coverage. Very few of these websites or blogs, however, are peer reviewed and so you must be wary of the accuracy of their content.

1.6.7 Vignette: Nontechnical Writing

The following is an example of nontechnical writing. It's an email to a friend or Facebook post concerning an accident in which I was involved:

> So yesterday I'm driving to the bank on Valley Road when this idiot runs a stop sign and T-bones my car. Thank God I wasn't hurt, but my side airbag deployed and that probably saved me. The car isn't totaled, but the passenger side door was crushed and there is a lot of other damage to it and they had to tow it. The cop who responded to the scene thought I was looking at $15,000 of damage or more. They towed my car to the body shop, but I haven't gotten an estimate to repair yet. The other guy was driving some old clunker and it barely had a scratch on it. Stupid jerk was probably drunk, but the cop didn't want to give him a breathalyzer test because I think he knew the guy.

Notice how the writing contains emotion, accusations, speculation, and blame, features that are not supposed to appear in technical writing.

1.6.8 Vignette: Technical Writing Sample

The following is an accident report written for an insurance company describing the same accident as in Vignette 1.6.7:

On October 1, 2017 at approximately 10 AM Driver A was proceeding by vehicle on Valley Road through a cross street. Driver B was proceeding through the stop sign at the cross street, colliding with the passenger side of the vehicle driven by Driver A. The driver side airbag for Driver A's vehicle deployed. There were no injuries. Driver A's vehicle needed to be towed.

Notice how this report avoids any emotional statements, speculation, or insults and focuses only on facts. You should always treat an accident as a technical report. Your lawyer will thank you for this advice if the situation ever escalates to a lawsuit.

1.7 Exercises

1.1 For the following, which could be considered to be technical writing, nontechnical writing, or both?

> An email to a friend about your new computer
>
> A complaint letter to the manufacturer of a robot vacuum cleaner
>
> A letter to your insurance company explaining how a disease affects you
>
> A letter to a government taxing agency in response to their inquiry about your tax return
>
> A written request to your work supervisor for a salary increase
>
> A letter to your local government authority requesting a zoning variance for a new pool

1.2 Consider the definition of "technical writing" given in the introduction. Based on this definition, is this book about "technical writing" technical writing itself, or not? Defend your answer. *Hint:* Where does this book fit in the taxonomy of Figure 1.1?

1.3 For a technical discipline of your choosing, identify a relevant:

> Blog
>
> Newsletter
>
> Magazine
>
> Journal
>
> Conference
>
> Make a note if you cannot find any of these

1.4 For the publications you identified in Exercise 1.3, find the publication's author guidelines (e.g., recommended paper length, aims and scope, submission instructions). These are generally found at the publication's website.

1.5 Using the same axes that were used in Figure 1.3, identify where the publications you discovered in Exercise 1.3 would appear on a similar graph.

1.6 Write a 300–400 word Facebook post describing your current place of residence.

1.7 Write a 300–400 word technical description of your current place of residence for an insurance company.

1.8 What are the main differences between your writings from Exercises 1.6 and 1.7?

1.9 Write a 50 word or less biosketch for yourself.

Endnotes

1. I was told a different version of this wordplay by members of the Royal Australian Air Force more than twenty years ago. In reference to diet, an Aussie would say, "A panda eats roots, shoots and leaves." But the panda is known to be quite amorous and a double entendre arises if you add a comma after "eats," that is, "A panda eats, roots, shoots and leaves"; the word "root" meaning the act of procreation. Thus, with one misplaced comma, you convert a harmless statement about panda dietary habits to pornography.

2. At least for the past decade or so, this has been the usual practice for technical periodicals. There are still many prior decades of important work not yet available in digital form.

References

Higham, N. J., *Handbook of Writing for the Mathematical Sciences,* The Society for Industrial and Applied Mathematics, Philadelphia, PA, 1998.

Montgomery, T. and Plung, D., A definition and illustrated taxonomy of technical writing, *Proc. of International Professional Communication Conference, 1988,* Seattle, WA, October 5–7, 1988, pp. 141–146.

NASA technical report, Mariner 1, available at http://nssdc.gsfc.nasa.gov/nmc /spacecraftDisplay.do?id=MARIN1, accessed October 2, 2017.

Truss, L., *Eats, Shoots & Leaves: The Zero Tolerance Approach to Punctuation,* Gotham, New York, 2004.

Wong, W. E., Li, X. and Laplante, Phillip, Be more familiar with our enemies and pave the way forward: A review of the roles bugs played in software failures, *Journal of Systems and Software,* 133, 68–94, 2017.

2

Technical Writing Basics

2.1 Introduction

Although it is an unfortunate stereotype, engineers and technical professionals are often regarded as poor communicators. When I was an engineering undergraduate student, a frequently heard insult was "I couldn't even spell 'engineer' and now I are one." Yet, most of the engineers and scientists I know are well read and articulate, and can write effectively.

I am a professor of neither English nor communications, and this chapter is not an introduction to basic writing principles such as verb tense consistency, avoiding run-on sentences, and proper transitions. I assume (really, hope) that you can write well enough. And I certainly won't bore you with the more esoteric subjects of introductory writing—for example, the "Harvard comma"[1]—I am just not that persnickety. For a great introduction to basic writing principles and for guidance on punctuation, grammar, and structure, I recommend *On Writing Well* [Zinsser 2006] and *The Elements of Style* [Strunk and White 2008]. These two books have profoundly influenced and improved my writing.

What I cover in this chapter are some basic rules that will help you improve any technical writing.

2.2 Structuring Your Writing

Whether you are writing procedures documents, manuals, reports, or books, it is conventional to organize your writing in a hierarchical fashion. Writing is hierarchical if it is arranged as a cascade of sections or chapters at a high level of abstraction which, in turn, are composed into sections of greater detail, and those sections into subsections, and so on, each at an increasing level of detail. The composition of writing in this fashion is used to help organize and convey ideas from high level to low level, that is, from abstract to concrete.

It is not uncommon to use a hierarchical numbering scheme to indicate the sections, subsections, and so forth as follows:

1 First-Level Heading

1.1 Second-Level Heading

1.1.1 *Third-Level Heading*

1.1.1.1 *Fourth-Level Heading*

For this book, I have used a variation of this scheme in which the first-level heading is a chapter, the second level heading is a section, and so on:

Chapter 1

1.1 Second-Level Heading

1.1.1 Third-Level Heading

1.1.1.1 *Fourth-Level Heading*

It is also possible to use different numeral types and fonts to denote higher and lower lever sections.

I. First-Level Heading

A. Second-Level Heading

1. *Third-Level Heading*

a. Fourth-Level Heading

It is uncommon to go further than fourth-level headings in most technical documentation.

Regardless of the numbering scheme chosen, it is good practice to balance the composition both externally and internally. External balance means that if you organize your writing into major sections, then subsections, and so on, then the relative numbers of each of these should be fairly uniform. Internal balance pertains to the granularity of the composition; that is, the relative length of the sections and subsections should be uniform.

For example, suppose you are writing some kind of systems requirements document and you are using the hierarchical numbering scheme so that 1, 2, 3, … represent top-level sections with high-level detail; 1.1, 1.2, 1.3, … represent second-level sections (more detail); and 1.1.1, 1.1.2, 1.1.3, … represent third-level sections with the most detail. You would expect that there would be more subsections at low-level detail than at high-level detail. So, for a hypothetical systems requirements document, if I listed the requirements at each level, the resulting shape should be a pyramid (see Figure 2.1a).

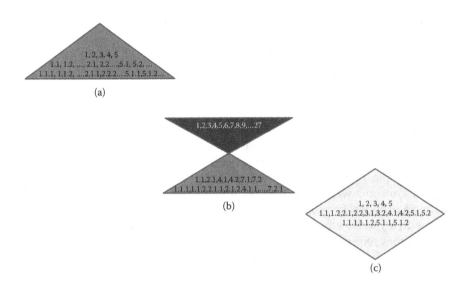

FIGURE 2.1
Pyramid (a), hourglass (b), and diamond-shaped (c) configurations for the heading levels in documentation. (From Laplante, P. A., *Requirements Engineering for Software and Systems, Third Edition*, Taylor & Francis, Boca Raton, FL, 2017.)

It is undesirable to have many high-level sections, a few subsections, and then many more at the lowest-level sections. That is, if you were to list the section numbers, the hourglass in Figure 2.1b would emerge. This situation indicates that there is missing mid-level detail. Finally, if there are a few high-level sections, lots of subsections, and fewer low-level sections, then there is a nonuniform level of discussion—too many mid-level details, but not enough low-level details. This situation can be visualized as the diamond in Figure 2.1c [Laplante 2017].

You may have to readjust the structure of your writing again and again as you write, edit, and rewrite in order to achieve the most desirable organization.

I tried to enforce a good internal and external structure in this text. The astute reader will notice, however, that I was not entirely successful. But in seeking an ideal organizational structure, your writing is not a failure if it doesn't resemble a perfect pyramid. You should simply try to achieve this result without sacrificing logical order and clarity.

There are other ways to organize your writing: a template, for example, as is the case in some grant proposals. You could also organize your document using official standards, industry conventions, or the publisher's requirements. However, it is undesirable to use a free-form or "stream of consciousness" style of writing, except during brainstorming. While it is possible, and even desirable, to draft this way, you will need to reorganize these thoughts into some logical structure in a final draft.

2.3 Positioning Your Writing

2.3.1 Know Your Audience

Before you begin writing, you must fully understand your intended readership and prepare to align your work accordingly. Different writing tones and approaches are needed for customer versus vendor, technical versus nontechnical personnel, and government agency versus other entities. Deciding on the audience early in the writing process will reduce later rewriting effort.

For example, nontechnical readers will require more explanation of introductory concepts while technical readers will want more detail. If you are writing a proposal or project report for a client, you will take a different tone than with a vendor who is working for you. Reports to government agencies and auditors require a deferential and clinical tone, whereas informal writing, such as in advertising, user manuals, and marketing copy, should be lighter in voice.

It is quite difficult to write to multiple audiences simultaneously. In such cases, the most formal tone should be adopted. For example, an incident report may be read by customers, government agencies, vendors, and management, but the report should be structured as the government agencies require.

2.3.2 Are You Talking to Me?

"First-person point of view" means writing from the point of view of the author. Thus, the words "I" and "me" will appear as noun subjects. Autobiographies and many novels are written this way. In some cases, technical writing can be in the first person, for example, in describing a series of events in which the writer is involved. In technical business communication, the first person is conventionally used.

"Second-person point of view" is used less often, and involves addressing the reader directly, that is, the writing is directed at "you," for example, in a procedures manual or user guide.

"Third-person point of view" is from the perspective of the author as an observer. This kind of writing uses personal nouns and pronouns such as Fred, Sue, him, and her. Certain technical writing should be in the third person because this point of view places the author in the position of an impartial, clinical observer. For example, the third person is preferred in accident or incident reports, descriptions of experiments and tests, and in project postmortem reports. Disclosure of research findings is almost always written in the third person.

It is not unusual for writing to have a mixture of first-, second-, and third-person elements. I wrote this book in the first and second person to make it intimate. Whatever voice you choose in your technical writing, be consistent with it throughout the document.

2.4 Choosing the Right Words

Conciseness and precision are two desirable characteristics of technical writing. The following sections describe how to achieve these qualities.

2.4.1 Conciseness

> Vigorous writing is concise. A sentence should contain no unnecessary words, a paragraph no unnecessary sentences, for the same reason that a drawing should have no unnecessary lines and a machine no unnecessary parts. This requires not that the writer make all the sentences short or avoid all detail and treat subjects only in outline, but that every word tell. [Strunk and White 2008]

Conciseness in writing is not easy to achieve. French Mathematician Blaise Pascal once wrote to a friend: "I have made this letter too long because I did not have the free time to make it shorter"[2] [Pascal 1656]. Pascal's sentiment acknowledges that it takes more effort to write concisely than to simply put down the first words that come to mind. But the effort is invaluable. Conciseness is achieved by finding ways to replace two or more words with fewer without losing meaning.

The power of word replacement is epitomized in the following, often-told tale about Benjamin Franklin. The story has many variations, often contradictory, but the lesson is always the same.[2]

Franklin was a scientist, inventor, statesman, philosopher, businessman, and a prolific author of technical and nontechnical writings. His many quotations from *Poor Richard's Almanac* (e.g., "A penny saved is a penny earned.") are widely used, even today, but his body of work, including books, pamphlets, and scientific papers, is astounding. To learn more about Franklin's life and writings, read his autobiography [Franklin 2010], which was originally published in French and translated into English after his death.

One version of the story is as follows: Franklin was asked to edit a first draft of the Declaration of Independence, written by Thomas Jefferson. Jefferson's draft was then reviewed by others, who made many substantive changes. Jefferson was not happy with the editing. To soothe his friend, Franklin recounted the story of a hatter, one John Thompson, who contracted to have a sign constructed for his shop. The proposed sign is given in Figure 2.2.

Because the sign maker charged by the letter, Thompson was looking to save money, so he showed his prototype sign to a group of friends. The first friend he consulted thought the word "Hatter" was extraneous, leading to the refinement shown in Figure 2.3. Another friend noted that the word "makes" might as well be omitted, because customers did not care who made the hats, leading to the version shown in Figure 2.4. The words "for ready money" seemed useless, noted another friend, because the journeyman hatter was in no position to sell on credit. The sign could then be reduced to the one shown in Figure 2.5. Because no one gives hats away, the word "sells" now

FIGURE 2.2
First draft of Thompson's sign.

FIGURE 2.3
Second draft of Thompson's sign.

seemed extraneous, another friend observed, leading to the version of the sign shown in Figure 2.6. Finally, Thompson himself realized that the word "hats" is clearly not needed because the meaning is conveyed in the picture. Therefore, the final version of the sign is shown in Figure 2.7.

This little story exquisitely illustrates how successive rounds of thoughtful editing and word replacement can foster conciseness without losing meaning. The power of using graphics is also evident.

Opportunities for obtaining conciseness through word replacement arise frequently, especially if, like me, you write in a conversational style. In conversational speech, there is usually no penalty for using excess words. In a dynamic exchange, we often can't select the one perfect word but can use two or more words to convey the same meaning. We can't and don't need

FIGURE 2.4
Third draft of Thompson's sign.

FIGURE 2.5
Fourth draft of Thompson's sign.

to edit our speech, and minor and even major mistakes are easily missed or forgiven. But these luxuries are not afforded when writing.

Reducing word count makes writing more precise. For example, look for ways to replace two words with one word of equivalent meaning. For example, "stretched thin" could be replaced by "overextended," "on the fly" by "contemporaneously," "write down" by "record," and "day to day" by "ordinary." Other examples are "as well as" which can be replaced by "and," "along those lines" by "correspondingly," and "came out" by "appeared." Table 2.1 gives several more examples of two or more word replacements in context.

Using too many words makes writing appear sloppy and amateurish. Do you agree that the reworked sentences in Table 2.1 are more authoritative?

FIGURE 2.6
Fifth draft of Thompson's sign.

FIGURE 2.7
Final version of Thompson's sign.

2.4.2 Precision and Hedging

When you can measure what you are speaking about, and express it in numbers, you know something about it; but when you cannot express it in numbers, your knowledge is of a meager and unsatisfactory kind; it may be the beginning of knowledge, but you have scarcely, in your thoughts, advanced to the stage of science.

William Thompson, Lord Kelvin, 1883

As Lord Kelvin noted, technical writing should be precise. Precision can be achieved using measurable quantities and avoiding vague modifiers such as

TABLE 2.1

Original and Improved Fragments with Replaced Words in Bold

Original Sentence Fragment	Reworked Sentence Fragment
The mechanism will require **a substantial amount of** redesign…	The mechanism will require **substantial** redesign…
The main housing mechanism **as well as** the linkage assembly must be thoroughly lubricated.	The main housing mechanism **and** the linkage assembly must be thoroughly lubricated.
There are many replacement parts **available out there.**	There are many replacement parts **available.**
Metal fatigue points **often can and do result** in major system failures.	Metal fatigue points **can result** in major system failures.
Once the diagnostic tests are finished without reporting a failure, you are ready to **fire up** the engine.	Once the diagnostic tests are finished without reporting a failure, you are ready to **start** the engine.
During the **give-and-take** session with the customer, the following concerns were raised:	During the customer **dialogue** session, the following concerns were raised:
A repair kit must **go along with** each unit sold.	A repair kit must **accompany** each unit sold.
Using the diagnostic system **installed on** the main system,	Using the diagnostic system **onboard** the main system,
The possible reasons for a subsystem failure are **many fold:**	The possible reasons for a subsystem failure are **numerous:**
After adding six quarts of reagent, the mixture will be **more clear.**	After adding six quarts of reagent, the mixture will be **clearer.**
Each unit will weigh, **more or less**, 70 kg.	Each unit will weigh **approximately** 70 kg.
The customer was **put off** by the latest test trials.	The customer was **unimpressed** by the latest test trials.
A **really strong** odor was noticeable.	A **powerful** odor was noticeable.
To **start off** the procedure,	To **begin** the procedure,
These facts, **taken as a whole**, indicate that a system failure was inevitable.	These facts **collectively** indicate that a system failure was inevitable.
One of the major actors in this disaster is speculating on a software **tie-in.**	One of the major actors in this disaster is speculating on a software **connection.**
The flexible strap is used for **tying in together**…	The flexible strap is used for **attaching**…
The vehicle must be able to operate in the presence of **very fast** currents.	The vehicle must be able to operate in the presence of **swift** currents.
The main software system should **work together** with the safety subsystem.	The main software system should **interoperate** with the safety subsystem.

"countless," "some," "approximately," "huge," "tiny," "microscopic," and so on. These vague words should be replaced with measurements.

For example, consider the following fragment from a description of some hydraulic system:

It is typical for some of the hydraulic fluid to escape from the housing, but this leakage should be minimal.

This statement is vague and gives an observer no real guidance to determine if a leak is significant. The following sentence is an improvement:

It is typical for some of the hydraulic fluid to escape from the housing, but this leakage should be less than one milliliter per day.

With this precise statement, an operator or inspector of the equipment involved could determine if an observed leakage is outside the norm.

Table 2.1 contains some imprecise words (e.g., "swift," "substantial"), which should be replaced with measurable quantities—if we had a reference measurement. For example, in Table 2.1, consider the following sentence fragment:

The possible reasons for a subsystem failure are numerous

It be better if we could write it as follows:

There are twelve possible causes for a subsystem failure

Of course, we can only write such a statement if we know that there are twelve causes of subsystem failure.

Suppose you are writing about a sample of 100 units of some manufactured part in a quality control report. You might write such sentence fragments as "only a few may be defective" or "a substantial amount should be within tolerance." But these statements are imprecise; it would be better to write, instead, that "less than ten may be defective," or "98 percent should be within tolerance," or to use some kind of range. Table 2.2 contains a set of imprecise words and their more precise replacements in both fractional form and as a range, where the base amount is 100.

Other forms of imprecision using fuzzy words can spoil your writing. For example, when you write that something happens "frequently," wouldn't it be better to give a range or an approximate rate. Table 2.3 contains some imprecise words and ranges of values that might replace those words.

TABLE 2.2

Replacing Imprecise Statements with Precise Numbers or Ranges

Imprecise	A Few	Some	Many	Most	All
Precise (fractional form)	1/100	1/20	1/2	90/100	100/100
Precise (range)	<10	10 to 20	21 to 50	51 to 99	100

TABLE 2.3

Ranges of Numbers that Could Replace Imprecise Rates

Imprecise	Always	Frequently	Often	Sometimes	Rarely	Never
Precise (as a ratio)	100%	99%–75%	75–50%	50–25%	25–1%	0%

So instead of "frequently," you might write "more than 75% of the time." Or, instead of "only a few may be defective", write, "<10 may be defective." Finally, "a substantial amount should be within tolerance" is improved as "98% shall be within tolerance." Providing a hard number or range allows for a reference point for future refinement of the information from a fuzzy number (or range) to a finer range or exact number.

You can't always replace hedging words with precision. Sometimes you can't offer much more than a guess. For example, consider a discussion of the likelihood of some event for which you have no prior statistics, and not even a probability model for constructing predictive statistics. This would be a case where hedging words are necessary.

Here is a short list of hedging words that should be replaced with a probability when possible:

- Likely
- Almost
- Probably
- Maybe
- Might be

I know I have used words from this list throughout this book, but only in the context of the narrative and where it would be impossible to create a realistic probability. For example, when I write that "Technical writing often contains equations," how could I make this statement more precise? Either I would have to review all the technical writing in the world to make this statement exact, or I would have to find and cite prior research that revealed more precisely the proportion of technical writing containing equations.

2.4.3 Universal and Existential Quantification

It is common in mathematical and engineering writings to use symbols to represent the concepts of universality and existence. One symbol, ∀, means "for all" or, equivalently, "always," "universally," and "completely," and many other variations on this theme. ∀ is called the universal quantifier. The other symbol, ∃, means "there exists" and is called the existential quantifier. Both symbols are powerful notations for making sweeping statements, but their use can be fraught with danger [Voas and Laplante 2010].

For example, sentences that require specification usually involve some universal quantification such as "All users shall be able to access the database." But is it really true that all users should be able to access the database? There may be some classes of users that should not access the database (e.g., new users), and the specification should reflect that reality. In this case, you would write, "All users, except new users, shall be able to access the database."

Related to "all" specifications are "never," "always," "none," and "each." But these can be formally equated to universal quantification [Laplante 2017]. In the previous sentence, for example, you could have written, "No new users shall be able to access the database."

Universal and existential quantification should be used with great caution. Unless "all" or "none" are demonstrably true, these words (and any of their equivalents) should be omitted.

2.4.4 Negatives

In technical writing, it is preferred to use the positive form of declarations, directions, and instructions rather than the negative. For example, instead of writing "do not include" it is better to write "exclude." Similarly, instead of "do not permit" use "forbid," and rather than "do not allow" write "disallow." The positive statements are more direct and promote the use of more precise language.[4] Similarly, in writing requirements specifications, you should use the positive form, rather than the negative form in structuring the requirements. That is, the requirement should be written using "shall" statements rather than "shall not" statements. Section 2.7.7 provides some examples of requirements statements in the positive form.

While it is desirable to avoid "shall not" requirements altogether, there are sometimes exceptions. Consider the following example:

The system shall permit access only to authorized users.

The requirement can be rewritten equivalently as:

The system shall not permit access to unauthorized users.

In this case, the first form is easier to understand because it does not involve a double negative. But consider the problem of designing test cases to prove that the requirement has been met. In the first form, the test cases have to involve subsets of authorized and unauthorized users. Then we need to decide if we are to use exhaustive testing, or equivalence class testing, or pairwise testing, and so on, for both authorized and unauthorized users. Conversely, in the second form, only a set of test cases involving one or more unauthorized user need to be created in order to show satisfaction of the requirement. We still have to decide on how to select a subset of the universe of unauthorized users, but the second form of the requirement simplifies testing significantly.

In some cases, there may be simply a psychological impact on stakeholders (particularly lawyers) in using the "shall not" form instead of the "shall" form of the requirement. Consider this example:

The system shall not harm humans.

This requirement seems more straightforward than the requirement:

The system shall do humans no harm.

In any case, you should write requirements using "shall" statements whenever possible, and when clarity is not sacrificed. But "shall not" requirements may occasionally be preferred to their "shall" equivalent form.

2.4.5 Vignette: Brake Retarder Prohibitions[5]

On a road I travel, there is a street sign declaring "End brake retarder prohibition." The meaning of this odd sign is hard to decipher and obscured by the curious use of a quadruple negative. But the sign is an interesting example of a "shall not" requirement (and ambiguous writing).

A brake retarder is a device used on large trucks to slow the engine, thus slowing the vehicle. Brake retarders are very noisy, and, as a result, many municipalities prohibit their use within city limits. Let's have a little fun with the sign. If we substitute the synonyms:

end → stop, brake → stop, retarder → stopper, prohibition → stopping

the sign translates to:

stop stop stopper stopping

which is a quadruple negative. In theory, we should be able to replace such a quadruple negative with an empty sign. The original sign has the paradoxical effect, however, of causing a positive action (i.e., to allow the application of brake retarders).

Of course, this whimsical discussion ignores the fact that the word "end" is a verb, "retarder" and "prohibition" are nouns, and in this context "brake" is an adjective, rendering a logical analysis of meaning more difficult. And the sign really does have the intended effect—truck drivers know what the sign means. But it should be clear that this particular sign can be interpreted multiple ways and is, therefore, ambiguous.

2.5 Avoiding Traps

There are certain writing mistakes that you must avoid in technical writing. Avoiding these traps will greatly improve the correctness and precision of your writing.

2.5.1 Clichés

I use clichés when I speak, which I regret. Worse, clichés have a habit of creeping into my writing. Clichés must be removed from any writing during editing, and the easiest way is to replace them with a word or phrase that has the same meaning. For example, "bite the bullet" means "sacrifice," and "in this day and age" can be replaced by "today" or "now." Table 2.4 contains some common clichés and their possible replacements.

A cliché is a lazy way to create an image. If you need to create such an image, do so through effective writing.

2.5.2 Anthropomorphic Writing

It is not appropriate to project human feelings, behaviors, or characteristics upon animals, inanimate objects, or systems. Such writing is called anthropomorphic, and it has a place in prose and literature but not in technical writing.

Anthropomorphic writing can be evidence that you have lost your objectivity. For example, consider a note in some test report in which you write:

> 8/23/2017, 3:11 PM EDT, upon running Test 3.1.2, the system failed miserably

Systems cannot be miserable—misery is a human condition. Of course, any person who was hoping that the system would succeed could be miserable upon observing failure, but that fact should not enter into the test report.

TABLE 2.4

Some Clichés and Possible Replacements

Cliché	Possible Replacement
Bite the bullet	Sacrifice
Don't mince words	Be precise
Drop in the bucket	Tiny
Feather in your cap	Accolade
Happier than a clam at high tide	Satisfied
In this day and age	Today
Ins and outs	Details
Nitty gritty	Details
No mean feat	Difficult
Only time will tell	Eventually
The world is your oyster	An opportunity arises
Whets the appetite	Invites

As an impartial observer, you should have simply written that "the system failed" and give details as to how.

Here is another example of anthropomorphic writing. It is often written that "the Internet is evil" (google it). But the Internet is neither evil nor good. The Internet is just a collection of computers, interconnections, protocols, services, and their supporting technologies developed by many people. Users of the Internet can be evil or good, but not the Internet itself.

2.5.3 Malapropisms

A malapropism is a word that sounds similar to an intended word but is logically wrong, often in some insidious way. Malapropisms are sometimes used for comic effect—Leo Gorcey of the old "Eastside Kids" comedies and comedian Norm Crosby were famous for their deliberate use of malapropisms. Apparently, the first use of malapropism comes from Richard Sheridan's 1775 play *The Rivals*, in which one "Mrs. Malaprop" misspeaks.

Malapropisms are particularly dangerous in technical writing because of subtle differences between the appropriate and inappropriate words that can have profound consequences. Table 2.5 contains a collection of malapropisms that one could plausibly find in some technical writing. I tried to make these examples funny.

Sometimes malapropisms occur when writing too quickly and selecting similar sounding words with the wrong meaning (see Section 2.5.4). Other times, malapropisms arise from translation errors from documents that originated in a different language.

Malapropisms can be insidious in technical writing because they may not be easy to detect. Notice that if you were to type the misuses from Table 2.5 into a word processor, the spelling and grammar checking feature might not catch all the errors.

2.5.4 Erroneous Heterographs

Two words are heterographs if they sound the same when pronounced but are spelled differently and have different meanings. For example, "compliment" and "complement" sound the same, but have different meanings—the former being praise and the latter meaning the opposite of something. Other heterographs that are often misused include:

course: coarse
facts: fax
fourth: forth
hire: higher
hole: whole

TABLE 2.5

Some Malapropisms and Sample Misuses

Malapropism	Example Misuse	Word Intended
Beet	"The rate at which the beets occur is called the beet frequency."	Beat
Contingency	"A contingency of technicians is needed for the repairs."	Contingent
Costumers	"We intend to keep the costumers happy by always changing with the times."	Customers
Destiny	"We expect reduced road surface life due to the increased destiny of the hot asphalt mix"	Density
Eros	"Any type 1 eros will be indicated by a flashing red indicator light."	Errors
Etymology	"Further testing of the pesticide's effectiveness would have to be conducted by the state's etymology lab."	Entomology
Examples	"In order to diagnose the condition, the physician must analyze several skin examples."	Samples
Floundered	"Due to poor sales, the launch of Product Alpha floundered."	Foundered
Ingratiate	"It is important that the final system will ingratiate a user-friendly interface with a high-performance database engine."	Integrate
Object-orientated	"The software should be designed in an object-orientated manner."	Object-oriented
Spoon	"The spoon rate of the fish can be increased 5% by elevating the average water temperature by 1°C."	Spawn
Typing point	"Increasing oscillations can lead to a typing point."	Tipping point
Wretched set	"You can easily assemble the kit using the wretched set provided."	Ratchet set

knew: new

real: reel

sheer: shear

their: there

two: too: to

The contraction "they're" (meaning they are) also sounds the same as "their" and "there" and is often used incorrectly in their place. Similarly, "it's" and "its" are frequently misused. "Its" is the possessive from of the pronoun "it," while "it's," is the contraction of "it is."

Using the wrong (but correct sounding) heterograph in writing can have important legal implications and also give a poor impression of the writer. So be very careful when writing in a conversational style (as I do) and especially if English is not your native language.

2.5.5 Opinion versus Fact

I have heard it said that "opinion is the lowest form of argument," but this sentiment is itself just opinion. Facts can sometimes be misleading. For example, academic studies involving a few students in a graduate course are not necessarily stronger evidence than real-world experience, particularly if the study contradicts industrial experience and other anecdotal evidence.

Suppose there is a published study illustrating the positive effects of alcohol on ten canaries in Lithuania. Can these results be imputed on all humans? Is the opinion that "you should not drink too much alcohol" less valid than the study? Even forgetting the many studies on the negative and positive effects of alcohol on humans, experience and common sense say that too much alcohol is bad for both humans and canaries.

Computer scientist and mathematician John von Neumann noted that "there is no sense in being precise when you don't even know what you're talking about" (www.great-quotes.com). And political economist Dr. Thomas Sowell stated that "It's bad enough that so many people believe things without any evidence. What is worse is that some people have no conception of evidence and regard facts as just someone else's opinion" (http://townhall .com).

While facts are essential in technical writing, there is a place for opinion. For example, informed opinion is valuable in user manuals, experience reports, and in describing applications of products or systems.

2.5.6 Acronyms, Domain-Specific Terms, and Jargon

Technical writing will contain various jargon, acronyms, and domain-specific terms. Do not assume that readers are familiar with these terms. Define any term that cannot be found in a standard dictionary or that has a meaning in context that is very different from a standard dictionary definition. It is conventional to spell out acronyms once (and only once) before using them. Jargon and domain-specific terms may also be defined before first use.

A glossary is a list of terms and their definitions, proper names (such as important agencies, organizations, or companies), and acronyms relating to the subject at hand. I have organized many glossaries—they are included in most of my books—and the two dictionaries that I edited were "super-sized" glossaries. Here is a short excerpt from the glossary of the second edition of my first book, *Easy PC Maintenance and Repair* [Laplante 1995], showing the format I like to use:

Baud: A data transmission rate in bits per second.

Binary expansion: Method of representing integer numbers using only combinations of bits.

Binary search: A divide-and-conquer technique in which the range of the search is halved each time.

Bit: The basic unit of computer storage. A bit of memory can be either a "1" or a "0."

Boards: See "cards."

Bootable disk: A disk that contains the boot portion of the operating system on it; also called a system disk.

Boot area: An area on a disk that contains a special code that allows the operating system to start.

Booting: The process of actually starting the computer's operating system.

Bootstrap code: Special code stored in the boot area of a disk that allows the operating system to start.

I realize that the terms in the above list are rather elementary, but at the time, home PCs and their jargon were new. Today, I would expect teenagers and even younger children to recognize most of the terms in the PC glossary.

Short glossaries can also be organized in tabular form and included anywhere in a document or on a technical website where jargon is used. Table 2.6 contains such a glossary for a Web-based tutorial that I wrote on open-source software.

TABLE 2.6

Open Source Software Terminology

Term	Definition
Committer	A member of an open source community who makes changes to the source code based on the rules of the community.
GPL	GNU public license, the most common open-source licensing model, which essentially says that any code that uses GPL licensed code must also be made available and freely distributed under the GPL. GPL is often derisively called a "viral" licensing model because of its self-propagation.
Open source software (OSS)	Software that is free for use or redistribution provided that the terms of a licensing agreement are followed. There are many different licensing models with varying degrees of rights and responsibilities.
Repository	A place where OSS can be found and where the associated communities are hosted. Typical repositories include SourceForge, RubyForge, and Freshmeat.
Software archeology	The study of a software program based on artifacts found in its open-source repository, such as developer logs, bug reporting information, and documentation.

Source: Adapted from Laplante, P., Open Source: The dark horse of software, *Computing Reviews*, July 2008, online at http://www.reviews.com/hottopic/hottopic_essay_09.cfm © 2008, ACM, Inc. With permission.

Here is some advice for organizing glossaries based on my experiences building them and building dictionaries:

If a word would be unfamiliar to someone outside your specialty, put it in the glossary.

If you are unsure if a word should go in the glossary, put it in the glossary.

Glossary entries that are defined by other jargon will require a new entry for each jargon term introduced.

Some glossary terms will need to be defined both in noun and verb form. For example, "interface" and "interfacing."

Use "see" and "see also" when a term in the glossary is similar to another word in the glossary.

Resist the temptation to put everything in the glossary (or you will end up embedding an English dictionary in your glossary).

Don't forget to consult the glossary of this book for more examples.

2.5.7 The Laziness of "Very"

Using the word "very" is a cheap way to try to amplify meaning. In the movie *Dead Poets Society* (1989) teacher, John Keating, played by Robin Williams eloquently stated: "avoid using the word very, because it's lazy. A man is not very tired. He is exhausted. Don't use very sad, use morose. Language was invented for one reason, boys—to woo women —and, in that endeavor, laziness will not do."

Any form of "very" paired with another word should be replaced by a single equivalent word. For example, "very large" could be replaced by "enormous" or "huge." "Very loud" could be replaced by "blaring," "very heavy" by "weighty," and "very clean" by "spotless." Table 2.7 lists several other examples.

TABLE 2.7

'Very'/Word Combinations and their Replacements

Very clear	Transparent
Very loud	Deafening
Very new	Pristine
Very often	Frequently
Very old	Ancient
Very quiet	Muted
Very strong	Mighty
Very weak	Feeble
Very fast	Sudden
Very simple	Basic
Very small	Tiny

Other vague qualifiers that are used in the same lazy manner as "very" include: almost, extremely, rarely, really, and massively. Word combinations using any of these should be replaced by more precise words, for example, "really often" by "frequently" and "extremely strong" by "mighty."

When possible it is always better to use measured quantities instead of using word replacement. For example, if referring to the height of a tree, "massive" is better than "very big," but writing "the tree is 250 feet tall" is best.

2.5.8 Other Pitfalls

The standard IEEE 21948 (2011) provides guidance for writing requirements, specifications, and other technical documentation. The standard suggests avoiding the following in your writing:

Comparative phrases, such as "better than" and "higher quality" since these are vague.

Indefinite pronouns, such as. "it", "this," and "that" since it can be unclear as to what they refer to.

Loopholes, for example, "if possible," "as appropriate," and "as applicable" since reduce the authority of the writing.

Nonverifiable terms such as "provide support," "but not limited to," "a number of" and "as a minimum."

Superlatives including "best," and "most" since these are often unverifiable

Subjective language including "user friendly," "easy to use," and "cost effective."

I would add that you should avoid using "useless" language. For, example, here is a selection of phrases that I have seen:

as a matter of fact

for what it is worth

all in all

that being said

in fact

that, in itself

I personally think that

These kinds of phrases add no information to the writing and should be avoided.

2.6 Making Your Technical Writing More Interesting

The term "technical writing" might cause you to think of several associated words, for example, "boring," "dry," "repetitive," "template," and "mechanical." None of these words has a positive connotation. It seems counterintuitive, but I contend that it is possible to make some technical writing interesting—even entertaining. Interesting writing is more likely to be read and retained, and it is therefore worth trying to achieve. Consider the following ways to make your writing more interesting.

2.6.1 Humor

I already mentioned that this book is a departure from traditional technical writing treatments, in that I have incorporated some humor and stories to hold your attention. But are there situations in conventional technical writing where a little humor is allowable, even desirable?

Yes, I think there are such situations. Consider, for example, the instructions found inside a build-it-yourself birdhouse kit. The assembly instructions need to be clear and unambiguous, but they can be charming. For example,

> Install the perch at a 90-degree angle to the face of the birdhouse, so your little feathered friends do not fall off.

You might want to scatter such light humor throughout the instruction manual to hold the reader's attention. There may be other kinds of technical writing where a little humor might be appropriate, for example:

- Technical reporting:
  - Manuals
  - Procedures
  - Planning documents
  - Bug reports
- Business communications:
  - Cover letters
  - Customer relations writing
  - Progress reports
  - Feasibility studies
- Scientific writing:
  - Books

- Magazines
- Conferences
- Newsletters
- Websites and blogs

This is not to suggest that humor is required—only that, with discretion and economy, it might be appropriate to use humor in some kinds of documents.

Of course, there are situations where humor is entirely inappropriate. For example, in a technical report on the performance of a new aircraft engine, it would be unwise to use humor. How do you think the following quip would be received?

> When the engine manifold temperature exceeds 500°C, you will be able to cook steaks on the manifold.

For the following types of technical publications, I advise against using humor:

- Technical reporting:
  - Specifications
  - Proposals
  - Facilities descriptions
  - Environmental impact statements
  - Safety analysis reports
  - Failure analysis reports
- Business communications:
  - Human resources communication
  - Trip reports
  - Administrative communications
  - Transmittal letters
  - Résumés
- Scientific writing:
- Journals

Humor also helps to organize and bond teams (see Chapter 10).

2.6.2 Vignette: The Joy of Spam

I tend to include some humor in my informal writings for scientific magazines, but only a light dose. But in one case I wrote a deliberately humorous treatise on email spam. In the paper, I discussed four levels of spam comprehension: fear, uncertainty, confidence, and knowledge, and if you isolate the first letters of those words you'll get a surprise. Another feature of the article was to list some of the randomly generated names created by spam bots. Some of these seemed quite funny to me, for example, Schmuck G. Deriding, Iroquoian L. Biscuit, Zirconium H. Coquetted, and Vealed C. Certitude [Laplante 2006].

Perhaps I went overboard on this article and I did not receive any praise for it. I suspect the piece may have been objectionable to some readers. I recommend you refrain from trying to be funny in anything that is published—confine your humorous intentions to private communications. Even then remember that private writings can become public.

2.6.3 Allegory

An allegory is a story that uses metaphors for real characters and events. The allegory is often used to disguise the real situation, or to make the meaning more universal across an archetype of problems. Generally, in technical writing I prefer directness to metaphor. But there may be instances where a metaphorical discussion is helpful, particularly in opinion essays, books, newsletters, websites, and blogs.

Here is an example to illustrate the use of allegory:

> When I was a boy my family frequented a Chinese restaurant near my home. We loved the food and the cheesy ambience, including hanging, colored, blowfish lamps. We ate at this restaurant about once every two months—as often as my mother could afford.
>
> One evening we went to the restaurant only to find it closed—condemned by the local Board of Health. We were confused and angry—why would they close such a fine establishment? Soon after, we learned through newspaper reports that the restaurant had been closed because they were serving cat and dog meat. As I grew older and had the opportunity to travel and eat at other Chinese restaurants around the world, I realized that the food at our favorite restaurant was actually terrible. We just didn't know it because we had no other Chinese food for comparison.

I used this allegory to convey to some students why I had reassigned their favorite electronics teacher to math courses—he was not teaching them modern techniques, but the students didn't know any better because he was the only electronics teacher they knew. I was prevented by a confidentiality agreement from telling the students the precise reason why the teacher was reassigned, so I attempted to convey the message through the allegory. This story is useful, however, because, as a metaphor, it can be used at any time to illustrate the point of "not knowing what you do not know."

Allegory is a useful technique in any kind of writing because it can help make boring material interesting, and because it can convey a point across a wide swath of applications. But be wary that the point of the allegory might be missed. For example, after telling the Chinese restaurant story, the students did not seem to understand my meaning.

2.7 The 5 Cs of Technical Writing

I previously noted that technical writing is characterized by two features—conciseness and precision. These are qualities of the writing style and of the author(s) of the work. But are there characteristics of good technical writing that go beyond stylistic excellence? It would be nice, for example, if there were a set of standards for good writing. Unfortunately, there are no such guidelines available.

There are some standards for specialized document types (e.g., IEEE 829–1998 for software testing [IEEE 1998]), but these generally provide only templates to help structure the documents. I know of no widely accepted, quality standards for technical writing. But we can draw guidance from requirements engineering, which uses a substantial amount of technical writing, and develop our own set of qualities of excellence for technical writing, which I call the "5 Cs."

2.7.1 Qualities of Good Writing

IEEE 830–1993 was a standard for specification documentation for "Requirements Engineering for Systems and Software" [IEEE 1993]. A System Requirements Specification is a prototypical kind of technical writing that is used in many industries to provide a high-level description of the functions and features of a proposed system. A requirements engineer is responsible for the activities needed to elicit requirements and then to translate those requirements into a written System Requirements Specification. Even though the IEEE Standard 830 was retired in 2011, it is still useful to us as a guide for technical writing. IEEE 830 proposed eight desirable qualities for SRS documents. These were:

1. Correct

2. Unambiguous

3. Complete

4. Consistent

5. Ranked for importance and/or stability

6. Verifiable
7. Modifiable
8. Traceable

Of these eight, five are quite relevant to any form of technical writing:

1. Correct
2. Unambiguous
3. Complete
4. Consistent
5. Modifiable

The other three qualities—ranked, verifiable, and traceable—may be relevant to most kinds of technical writing beyond a requirements specification, but I believe that only the set of five is relevant to *any* kind of technical writing. Making equivalent word substitutions for quality 2 and quality 5 leads to a list of "5 Cs" of technical writing:

1. Correct
2. Clear
3. Complete
4. Consistent
5. Changeable

Precision, one of the two special qualities of technical writing discussed in Chapter 1, is a combination of clarity and correctness. I'll discuss each of these five characteristics further.

2.7.2 Correctness

Correctness means that the information in the written artifact is grammatically and technically correct. For example,

> The automobile weight shall be no greater than 200 kilograms.

is clearly incorrect ("2000 kilograms" was intended). The next sentence,

> The automobile shall weight shall be no greater than 2000 kilograms.

is grammatically incorrect because of the extraneous "shall."

Testing for correctness in any kind of writing is difficult. You can use grammar and spell checkers, but they are not 100% accurate, and they cannot check technical correctness or the correctness of ideas. No spell checker will flag "200 kilometers" as incorrect because there is no spelling error, only a logic error. Review by one or more persons, or via group reviews, can increase the correctness of any writing. I discuss review and revision further in the next chapter.

2.7.3 Clarity

Clarity (or unambiguousness) means that each sentence, related groups of sentences, or related sections of the written document can have only one interpretation. Let me illustrate clarity by contradiction with an example.

In an automobile that I own, an indicator light is displayed when certain exceptional conditions occur.[6] These conditions include poor fuel quality, fuel cap not tightened properly, and other fuel-related faults. According to the user's manual, if the cause of the problem is relatively minor, such as the fuel cap not being tightened, the system will reset the light upon:

"removing the exceptional condition followed by three consecutive error-free cold starts. A cold start is defined as a start-up that has not been preceded by another engine start-up in the last eight hours, followed by several minutes of either highway or city driving."[7]

Isn't this confusing? If you wait eight hours from the previous start-up, then start the engine to drive somewhere, you have to wait at least eight hours to start up and drive back to your origin. If you have any warm start before three consecutive cold starts, the sequence must begin again. It seems that the only possible way to satisfy this condition is to drive somewhere, wait there for eight hours, and then drive back to the origin three times in a row. Or, you can drive around for a while, return to the origin, wait eight hours, then repeat this sequence two more times. The logic is very hard to follow and, in fact, after one month of trying, I could not get the light to reset without disconnecting and reconnecting the battery.

I hope examples of clarity in writing are abound in this book.

2.7.4 Completeness

A technical document is complete if there is no missing "relevant" or "important" information. Of course, "relevant" and "important" are relative terms—relative to the reader's needs, that is.

Completeness is a difficult quality to prove for any writing. How do you know when something is missing? The most powerful technique for reducing incompleteness is to have as many persons read the material as possible. In some technical writing, where subsequent versions of the document are expected, you can keep track of missing information as it is identified and then add that material to the next version of the document.

2.7.5 Consistency

The consistency of a document can take two forms: internal and external. Internal consistency means that one part of the document does not contradict another part. External consistency means that the document is in agreement with all other applicable documents and standards.

Internal and external consistency can be checked through reviews and can be repaired in subsequent versions of the document as inconsistencies are identified.

2.7.6 Changeability

A document is changeable if the structure of the document will readily yield to modification. Usually this means that the document is numbered, stored in a convenient electronic format, and compatible with common document processing and configuration tools.

It is obvious why changeability is an important quality of any technical document—the contents will change as errors and omissions are identified. Ease of modification will also reduce costs, assist in meeting schedules, and facilitate communications with respect to any projects that are associated with the technical documentation in question. Reviews and inspections are the most obvious way to assess a document's modifiability.

2.7.7 An Example

How technical writing may be evaluated along the 5 Cs can be illustrated by a simple example.[8] This example is an excerpt from a systems specification for a computer-automated or "smart" home entertainment system. The following excerpt refers to an intelligent solution for recording shows or movies that play through the television.

8.1.1	The system shall allow user to record any show on television.
8.1.2	The system shall present a Web interface with a grid listing similar to the *TV Guide* book for users to select shows to record.
8.1.3	The system shall allow user to record a minimum of two (2) television shows simultaneously.
8.1.4	The system shall make storage for recorded shows expandable.
8.1.5	The system shall free storage space as needed by first in, first out (FIFO) or some other defined priority schedule.
8.1.6	The system shall provide search feature to search through television shows to select which one to record.
8.1.7	The system shall provide user the ability to record all occurrences of a specified show.
8.1.8	The system shall provide user the ability to record only new instances of a specified show.

8.1.9   The system shall provide telephone menu options for customer to dial in and select channel, time, and duration to record.

8.1.10   The system shall present users option to select quality for recording.

8.1.11   The system shall present user option to not automatically overwrite the television recoding.

8.1.12   The system shall give user the option to only store X number of episodes from a certain series at a time.

8.1.13   The system may skip commercials when the system is able to detect the commercial.

8.1.14   The system shall monitor storage space for future recordings.

8.1.15   The system shall send notification when resources get low enough where recordings will be overwritten.

8.1.16   The system shall permit users to not automatically delete a show or a series.

8.1.17   The system shall not record any new shows if there is space available for recovery.

8.1.18   The system shall send notifications to users if there is no longer space available to record new shows.

Line numbering is typically used in SRS documents to help in referencing specific requirements later, and this feature is helpful to us now.

So, let's evaluate this sample of technical writing according to the 5 Cs:

Correct. The requirements appear to be consistent with common understanding of television recording works in a home environment. But it would be hard to be certain if this set is correct without further analysis and the use of focus groups and experts.

Clear. There are quite a few unclear or ambiguous phrases in these requirements that are revealed by asking, "How do I test this?" For example, 8.1.1 says that the system shall record "any show on television." There are many problems with this statement: Any show on U.S. television or around the world? What about pay-per-view, and so on? Requirement 8.1.4 says the system shall "make storage" for recorded shows expandable. What does "make storage" mean? Allocation? How much expansion should be allowed? Requirement 8.1.6 talks about searching—but based on what? Program name, time it was recorded, subject? Requirement 8.1.10 describes selecting "quality for recording." What does "quality" mean? Requirement 8.1.12 mentions storing "X episodes." Is "X" a natural number, an integer, a complex number, a quaternion? There are other ambiguities here. With no disrespect to the author, this list needs to be improved.

Complete. We can't really say if this set of requirements is incomplete without seeing the rest of the specification and without a thorough peer review.

Consistent. We cannot be sure about the consistency of this writing, and there is truly no way to guarantee consistency for a document written in English (or any other natural language for that matter). Only technical documentation written using formal methods (which essentially look like some combination of a programming language and mathematics) can be rigorously evaluated for completeness. In this situation, however, an informal reading is probably adequate. An informal reading of the requirements does not reveal any obvious inconsistencies.

Changeable. There are no obvious challenges to modifying and maintaining these requirements. The use of an appropriate change management tool to track the changes, and their sources and dates, would be needed.

For any technical writing, the best way to establish these desirable qualities is through peer review. There are some automated tools that might help establish the presence (or absence) of these qualities. For example, spell checkers, grammar checkers, and writing-level indicators might help establish whether or not the document is ambiguous.

2.8 Referencing

2.8.1 Choose the Right References

When selecting references to use in a technical paper or in any kind of technical writing, the use of good judgment is essential. For example, Web-based references can be of dubious origin, so you should not use too many of these (see Section 2.8.2). References that are too old, say more than ten years old for rapidly changing technologies, are also inappropriate.

Are the right people being cited in the references, for example, well-known experts versus unknown persons? Are the classic references in the field being cited, that is, those references that are considered the definitive or pioneering works in the field? Are you using enough references? I can't give you a specific count, but the authority of any technical publication will be challenged if too few references are given. You can also over-reference, that is, have too many references—particularly if the references are duplicative in some way (e.g., two similar works by the same author).

Your reference list should not include a preponderance of books (because books age quickly), conferences (because these are not fully vetted), journal articles (because these are often too theoretical), or Web references (because these do not have the same vetting and cognitive authority as the other types of references). Indeed, a balance of these reference types is most desirable. Your reference list says a lot about your familiarity with the subject matter, so use the right mix of references.

2.8.2 Web References

When my son was twelve years old, he posted an entry to Wikipedia ("crapware") that remained relatively unchanged for several months. This experience amplified my distrust of unattributed Web-based sources.

In fact, research by Goldman [2010] showed that most savvy readers are less likely to trust a nonymously authored Web sources (Wiki-style authoring) than named-expert produced (conventionally authored) content. Goldman found that conventional authoring is "better" than Wiki-style authoring, as the former is more credible and accurate, and the authors are more knowledgeable. Goldman's other findings include:

Readers believed that conventional-style authoring is more authoritative.

Factual articles were perceived to be better than persuasive articles.

More frequent Web users found Wiki-style articles less credible, less accurate, and had less knowledgeable author(s).

Men were more critical than women of articles created through Wiki-style authoring compared with those created by a conventional authoring method.

Women also found the conventionally authored material to be more credible, but the distinction was not as sharp as it was for men.

Goldman offers advice for the writer using Web-based references:

Identify the authors, if possible.

Show usage data, if it is available.

Know your audience with respect to gender.

These rules are even more important in persuasive writing, or when writing in an area in which you are unfamiliar [Goldman 2010].

When using anonymously authored sources, you must be very cautious. I do not recommend using such references unless the information comes from a reputable website (e.g., NASA, the American Medical Association, or a university). Even in these cases, beware of political agenda bias.

2.8.3 Reference Styles

There are many ways to represent the references or bibliography for technical documentation. By "represent" I mean the manner in which you list the author or authors, venue of publication, and other pertinent facts that allow a reader to locate that reference.

Every outside source that you use must be referenced, including websites, interviews, television programs, conference papers, books, journal articles, and so on. If you use information from any source literally, then it must appear in quotations, or offset somehow in the case of very long quotes. Even if you are only using ideas from another source but not literally quoting, you must reference that information. In the case of very long quotations, be sure to check if you need permission to use these (see Section 3.6.1).

There are many standard referencing styles. Scholarly and professional organizations publish these styles—for example, the American Psychological Association (APA), the Modern Language Association (MLA), the American Institute of Physics (AIP), and in the case of technical writing, The Institute of Electrical and Electronics Engineers (IEEE). Each of these referencing styles is fine. Use the style that is required by your employer, industry, client, professor, or publisher. My publisher, Taylor & Francis, uses a version of the Chicago Manual of Style (CMS), so that is what I used in this book.

I don't like to get hung up on referencing style. I keep track of references and citations from the start of my writing, but formatting the reference styles is unpleasant, and it is one of the last things that I do. It is often convenient to use bibliographic database software to organize and help format your references. Common software tools for this purpose include ProCite, EndNote, and RefMan.

For a thorough discussion of referencing for technical documentation, see Hodges et al. [1999] or Turabian et al. [2007].

2.9 Exercises

2.1 Which of the structures in Figure 2.1 most closely resembles the hierarchy of:

   a. Chapter 2

   b. Chapter 8

   c. The entire book

2.2 For the following negative statements, replace the combination with a single word:

   a. very painful

   b. very rich

   c. very strong

2.3 Write five sentences from some hypothetical technical document, each containing a cliché. Rewrite those sentences to remove the cliché.

2.4 Create a set of five malapropisms that could plausibly occur in some technical documentation. Try to be funny. Rewrite those sentences to remove the malapropism.

2.5 For the following "very"/word combination replace it with a single, more precise word:

   a. very short

   b. very pure

   c. very hard

2.6 Find five instances where I have used hedging or imprecise words in this book and then rewrite those sentences to remove the imprecision.

2.7 For Section 2.4.5 of this chapter, conduct a "5 Cs" analysis (see Section 2.7.7).

2.8 Some of the reworked sentences in Table 2.1 can be improved. Use word substitutions or rearrangement to make as many improvements to Table 2.1 as you can.

2.9 Rewrite the automobile user manual instructions more clearly regarding resetting the indicator light (see Section 2.7.3).

2.10 Rewrite the partial specification for the smart home entertainment system in Section 2.7.7 (Requirements 8.1.1 through 8.1.18) so that it more closely comports with the 5 Cs.

Endnotes

1. That is, whether the last item in a sequence should be written as "lock, stock and barrel" or if it should have the "Harvard comma," written "lock, stock, and barrel."
2. Similar quotes, but much later, are attributed to others, including Benjamin Franklin and Irish playwright George Bernard Shaw. Mark Twain (Samuel Clemens) several times made the same point about preparation for public speaking.
3. I discovered this vignette in a wonderful book on software refactoring [Kerievsky 2004], but the version I will recount is adapted from [Williamsburg 2017]. The lesson within the lesson here is to be careful to double-check your sources.
4. The following discussion is excerpted and adapted from Laplante [2017] with permission.
5. The following discussion is excerpted and adapted from Laplante [2017] with permission.
6. A variation of this example appeared in my requirements engineering book [Laplante 2017].
7. I have altered this writing to avoid exposing the identity of the vehicle.
8. This example is adapted from my requirements engineering book [Laplante 2017].

References

Dead Poets Society, Touchstone Pictures/Silver Screen Partners I, 1989.

Franklin, B., *Autobiography of Benjamin Franklin*, CreateSpace Publishers, www.create space.com, 2010.

Goldman, J., The Cognitive Authority of Collective Intelligence, PhD dissertation, Drexel University, Philadelphia, PA, May 28, 2010.

Hodges, J. C., Horner, W. B., Webb, S. S., and Miller, R. K., *Harbrace College Handbook: With 1998 MLA Style Manual*, Harcourt Brace College Publishers, Fort Worth, TX, 1999.

IEEE Std 21948: Systems and software engineering—Life cycle processes—Requirements engineering, IEEE, Piscataway, NJ, 2011.

IEEE Std 829-1998, IEEE Standard for Software Test Documentation, Institute of Electrical and Electronics Engineers, Piscataway, NJ, 1998.

IEEE Std 830-1993, IEEE Recommended Practice for Software Requirements Specifications, Institute of Electrical and Electronics Engineers, Piscataway, NJ, 1993.

Kerievsky, J., *Refactoring to Patterns*, Addison-Wesley, Boston, MA, 2004.

Laplante, P. A., *Easy PC Maintenance and Repair, Second Edition*, Windcrest/McGraw-Hill, Blue Ridge Summit, PA, 1995.

Laplante, P.A., *The Joy of Spam*. Queue, vol. 4, no. 9, 2006, pp. 54–56.

Laplante, P. A., *Requirements Engineering for Software and Systems*, Third Edition, Taylor & Francis, Boca Raton, FL, 2017.

Laplante, P., Open source: The dark horse of software, *Computing Reviews*, July 2008, online at http://www.computingreviews.com/hottopic/hottopic_essay_09.cfm], accessed December 13, 2017.

Pascal, B., *Lettres Provinciales* (1656–1657), No. 16.

Sowell, T., *Random Thoughts*, Townhall online, http://townhall.com/columnists/thom assowell/2002/07/05/random_thoughts, accessed December 13, 2017.

Strunk, W. and White, E. B., *The Elements of Style: 50th Anniversary Edition*, Longman, London, 2008.

Turabian, K., Booth, W. C., Colomb, G. G., and Williams, J. M., *A Manual for Writers of Research Papers, Theses, and Dissertations, Seventh Edition: Chicago Style for Students and Researchers (Chicago guides to Writing, Editing, and Publishing)*, University of Chicago Press, Chicago, IL, 2007.

Voas, J. and Laplante, P., End brake retarder prohibitions: Defining "shall not" requirements effectively, *IT Professional*, 12(3), 46–52, 2010.

Williamsburg History Center, Thomas Jefferson: An Anecdote of Doctor Franklin, http://www.history.org/almanack/resources/jeffersonanecdote.cfm, accessed December 13, 2017.

Zinsser, W., *On Writing Well, 30th Anniversary Edition: The Classic Guide to Writing Nonfiction*, Harper Books, New York, 2006.

3

The Writing Process

3.1 Introduction

Winston Churchill, a prolific writer of historical and political material, won the Nobel Prize for literature in 1953. He characteristically wrote while standing and would continuously edit and re-edit his work until he could do no more. According to Churchill,

> Writing a book is an adventure. To begin with, it is a toy and an amusement; then it becomes a mistress, and then it becomes a master, and then a tyrant. The last phase is that just as you are about to be reconciled to your servitude, you kill the monster, and fling him out to the public. [Manchester 1988]

Contrast Churchill's view on writing with that of science fiction writer Isaac Asimov, who authored more than 500 books:

> Thinking is the activity I love best, and writing to me is simply thinking through my fingers. I can write up to 18 hours a day. Typing 90 words a minute, I've done better than 50 pages a day. Nothing interferes with my concentration. You could put an orgy in my office and I wouldn't look up—well, maybe once. [Barbato and Furlich 2000]

Finally, author and journalist George Orwell reflected that

> All writers are vain, selfish and lazy. [Orwell 1947]

These observations from some of the world's greatest English-language writers show that every writer is different. My own experiences in writing, particularly books, echo Churchill's much more than Asimov's and, I hope, Orwell's. The trick to successful writing is to structure the writing project so that it is fun, and then the result will be much better than if the project is tedious.

If you have an interest in writing professionally, here is something to consider. All writers get writer's block, and some authors even find writing to be excruciating at times (e.g., J.D. Salinger and Truman Capote). But if you wish to be a writer, you should write every day, even if it is only for a few minutes.

3.2 The Traditional Writing Process

Take a look at any high school freshman writing text, and you are likely to see a five-phase writing process: brainstorming, drafting, revising, editing, and publishing (see Figure 3.1). This characterization is a reasonable one; but like any linear, sequential model, the distinction between these phases is not always very clear. For example, you don't know exactly when you have transitioned from brainstorming to drafting. You also might track back from one or more phases, and the cycle may be repeated in its entirety, say, when a major revision occurs.

Figure 3.1 reminds me of a saying that was popular with computer programmers many years ago, "Make it run, make it right, make it good." What was meant was that you needed to get some code written and get it so that it would compile and run and actually produce a result ("make it run"). Then you would make sure that the results produced were correct ("make it right"). Finally, you tried to make the program efficient and the code more understandable ("make it good"). While this approach to writing software isn't advisable, it is an appropriate process for technical writing; that is, get something down, organize it, clean it up, make it better, and then publish it.

Although these phases are depicted in Figure 3.1 as equal in duration, this is not really the case. My experience with most kinds of technical writing is that I spend about 20% of the time brainstorming, 40% drafting, 25% revising, 10% editing, and 5% publishing. These proportions will differ, however, depending on the technical document type, the writer, and the venue for the publication. Let me describe in more detail my experience using the process shown in Figure 3.1.

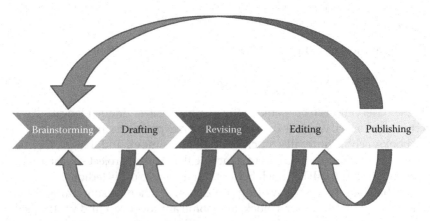

FIGURE 3.1
The writing process.

3.2.1 Brainstorming

Brainstorming, sometimes called "prewriting," is the process of recording your ideas on paper. In creative writing, this process can be very free-form and can include any ideas, no matter how crazy they may seem. In technical settings, this kind of brainstorming usually takes place in product research and development, often in project teams. But if you are writing some kind of technical report, or a requirements specification, brainstorming has nothing to do with creativity. It simply means that you need to start writing snippets of text and worry about organizing them later. During this phase, you should also consider who your audience will be, such as a customer, management, a vendor, or a government agency.

When in the brainstorming phase, I keep a file, either electronic or paper, with various ideas, quotes, pointers to other resources, and so on. For example, here is an initial brainstorming list for an article I wanted to write on how software engineering as a profession may be perceived differently in different countries.

Hofstede cultural dimensions

- Check for previous work

Outliers: See Gladwell's discussion of the Korean Airlines crashes

Story about a guy I met at the MTA meeting

PE PAKS survey—international questions

Offshoring

Third-party and externally furnished components of dubious origin

English as a second language for U.S. engineers

At this point, the list of ideas probably means very little to you, but it's a tickler file, a set of handles for me. Each point is something to be explained, explored, or expanded as the brainstorming continues. This idea file changed, grew, and matured until I had enough information to begin writing small portions of text. Eventually, this loose collection of ideas evolved into a well-developed paper that appeared in my regular column in *IT Professional* [Laplante 2010].

If you are visually oriented, you can use concept (or mind) maps to organize your ideas during brainstorming. A concept map is a hierarchical representation of ideas showing the relationships from a central concept to various subconcepts and sub-subconcepts. Figure 3.2 is a concept map that I might use to begin to write a proposal to the vice president for a new engineering department that I would like to create.

You will immediately notice that many elements are missing from the "proposal" in Figure 3.2. For example, what is the schedule for the new department? How is this department different from other departments? What are the downsides of creating a new department? These kinds of

FIGURE 3.2
A brainstorming concept map for a proposal to create a new technical department.

questions highlight one of the strengths of visual brainstorming—they can help identify gaps in coverage. Once you have identified those gaps, you can update the concept map diagram. When you are satisfied that the diagram is substantially populated, you can use the nodes in the diagram as section headings, or subsection headings, and begin the draft writing phase.

I usually advance from the brainstorming to the drafting phase when I feel that I have about 90% of the section headings and ideas listed. But there is no set time when you should leave brainstorming for drafting. You can, in fact, continue to brainstorm new ideas throughout the entire writing process.

You can also brainstorm in advance for future writing projects that you might not be ready to work on simply by keeping a paper or electronic journal of interesting ideas and questions as they occur to you.

3.2.2 Drafting

For me, the drafting process is the hardest part, and it usually takes the longest time. During this phase I have to write English prose, whereas in brainstorming I only have to record spontaneous ideas.

For the drafting process, I have to begin writing full sentences, even paragraphs, around each of the topical headings that I created during brainstorming. Then I write the "connective tissue" between these elements. I don't worry about grammar or spelling yet—the spell checker catches most spelling errors anyway.

At some point while writing, I usually have to rearrange sentences so that they make sense and provide "glue" or transitions between ideas that might not be obviously connected. If things go well, I will produce something that resembles the final document. If things do not go well, I may have to throw away much of the work and restart, although, thankfully, this scenario rarely happens.

The above advice won't make writing easy, only easier. The drafting phase is the stage where you have to just bear down, be fearless, and write, write, write. When you think you have written enough so that your document is about 90% complete, you can start revising.

3.2.3 Revising

Once I have produced a complete-looking document, it is time to make the document "good" through revision. For a fresh perspective, I usually wait a few days between the time I finish my first draft and the time I start my first revision pass.

Remember Pascal's quote from Chapter 2, noting that writing a short letter takes more time than writing a longer one? You should also expect that a revised manuscript will be much shorter than an early draft. To see the effects of a revision cycle, consider a marked-up page from an early draft of the foreword to this book, shown in Figure 3.3. I don't expect that you can read my handwriting in the figure (I often can't read my own handwriting), but you can see how heavily I have marked up this draft.

The result of making the changes shown in Figure 3.3, along with other contemporaneous edits, is shown in Figure 3.4. The version of the foreword shown in Figure 3.4 is not the final version. Several more rounds of revisions and a final edit by a copyeditor were needed prior to publication.

It is also important to have others review your work during the revising process. I always advise having at least two people review technical writing: one who is familiar with the subject matter and one who is nontechnical but who is otherwise an excellent writer. The nontechnical person often finds logic errors that the technical reviewer does not. My wife, who has a PhD in Nursing, reviews all of my writing. In exchange, I review her work.

If you are going to ask others to review your writing, be prepared for criticism. Recall Benjamin Franklin's anecdote from Chapter 2 about the hatter, John Thompson. Franklin used the Thompson anecdote as a metaphor for why he did not submit his writings to critique by committee—he did not like criticism [Franklin 2010].

It is hard to know when to stop revising. It is very rare that I feel that a document I am writing cannot improve with further revision. I usually stop revising when I run out of time and have to finalize the document. If I find that I am making revisions and then undoing those same revisions, then this usually signals that I should stop revising.

3.2.4 Editing

A classical editorial process is a more formal one involving editing by a professional copyeditor. During editing, the fine points of presentation, grammar, and formatting are perfected. The writing may also be made more precise through word substitution and reduction. The copyeditor will also identify any inconsistencies in logic and will check the references and citations. In short, the copyeditor will apply the 5 Cs.

This book received an excellent copyedit (see "Acknowledgments"). The copyeditor's job was made more challenging because of the deliberate mistakes used to illustrate points, which had to be left unaltered. Writing submitted to a scientific journal or magazine will also be subjected to copyediting.

If you are not submitting your writing for formal publication, or there is not another person playing the role of copyeditor for you, then you must become your own copyeditor. A copyeditor has to pay great attention to

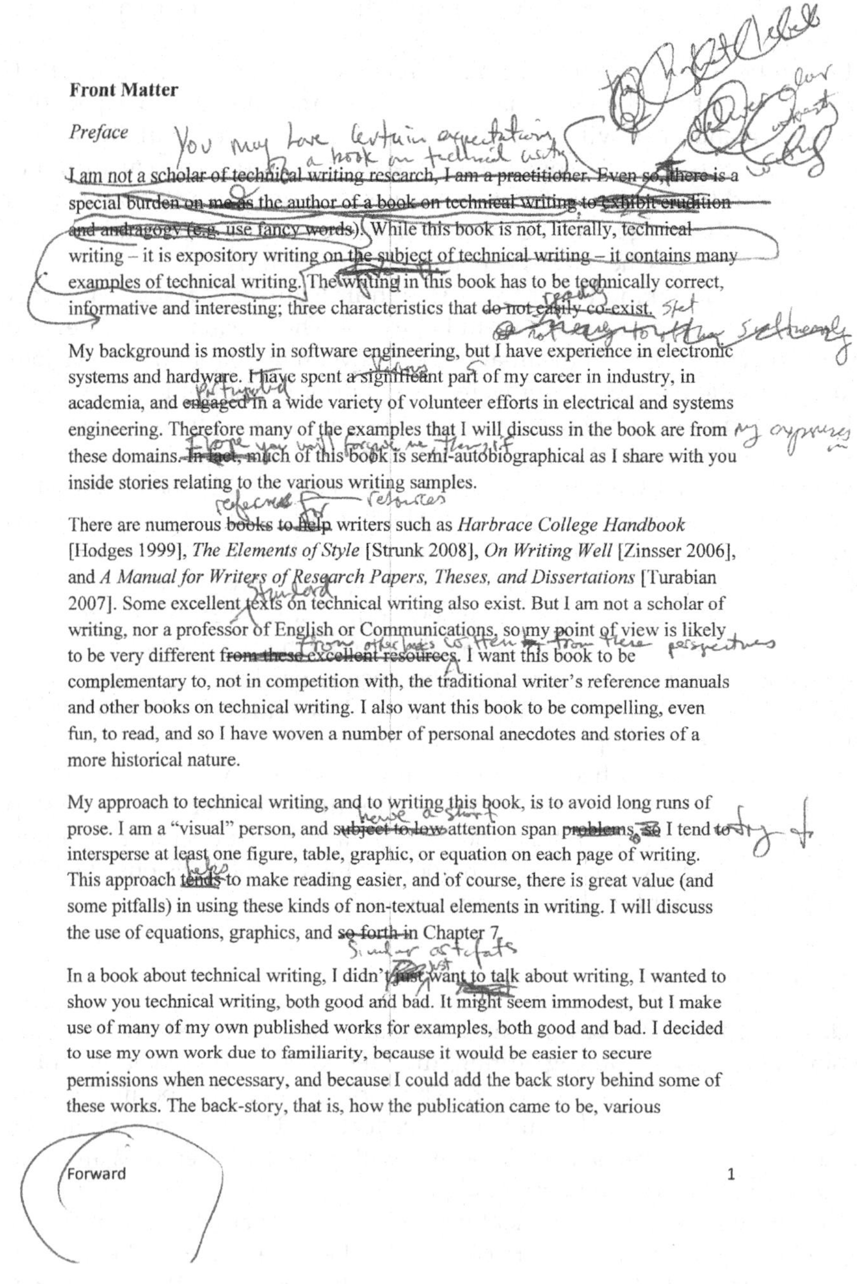

Front Matter

Preface

I am not a scholar of technical writing research, I am a practitioner. Even so, there is a special burden on me as the author of a book on technical writing to exhibit erudition and andragogy (e.g. use fancy words). While this book is not, literally, technical writing – it is expository writing on the subject of technical writing – it contains many examples of technical writing. The writing in this book has to be technically correct, informative and interesting; three characteristics that do not easily co-exist.

My background is mostly in software engineering, but I have experience in electronic systems and hardware. I have spent a significant part of my career in industry, in academia, and engaged in a wide variety of volunteer efforts in electrical and systems engineering. Therefore many of the examples that I will discuss in the book are from these domains. In fact, much of this book is semi-autobiographical as I share with you inside stories relating to the various writing samples.

There are numerous books to help writers such as *Harbrace College Handbook* [Hodges 1999], *The Elements of Style* [Strunk 2008], *On Writing Well* [Zinsser 2006], and *A Manual for Writers of Research Papers, Theses, and Dissertations* [Turabian 2007]. Some excellent texts on technical writing also exist. But I am not a scholar of writing, nor a professor of English or Communications, so my point of view is likely to be very different from these excellent resources. I want this book to be complementary to, not in competition with, the traditional writer's reference manuals and other books on technical writing. I also want this book to be compelling, even fun, to read, and so I have woven a number of personal anecdotes and stories of a more historical nature.

My approach to technical writing, and to writing this book, is to avoid long runs of prose. I am a "visual" person, and subject to low attention span problems, so I tend to interspersed at least one figure, table, graphic, or equation on each page of writing. This approach tends to make reading easier, and of course, there is great value (and some pitfalls) in using these kinds of non-textual elements in writing. I will discuss the use of equations, graphics, and so forth in Chapter 7.

In a book about technical writing, I didn't just want to talk about writing, I wanted to show you technical writing, both good and bad. It might seem immodest, but I make use of many of my own published works for examples, both good and bad. I decided to use my own work due to familiarity, because it would be easier to secure permissions when necessary, and because I could add the back story behind some of these works. The back-story, that is, how the publication came to be, various

Forward   1

FIGURE 3.3
Marked-up early draft of front matter for this book.

Front Matter

Preface

You will have certain expectations of a book about technical writing. The writing has to be technically correct, informative and interesting; three characteristics that do not readily co-exist. This book is not, literally, technical writing—it is expository writing on the subject of technical writing. But this book contains many examples of technical writing.

My background is mostly in software engineering, but I have experience in electronic systems and hardware. I have spent various parts of my career in industry, in academia, and participated in a wide variety of volunteer efforts in electrical and systems engineering. Therefore many of the examples that I will discuss in the book are from my experiences in these domains. I hope you will forgive me, then, if much of this book is semi-autobiographical as I share with you inside stories relating to the various writing samples.

There are numerous resources for writers such as *Harbrace College Handbook* [Hodges 1999], *The Elements of Style* [Strunk 2008], *On Writing Well* [Zinsser 2006], and *A Manual for Writers of Research Papers, Theses, and Dissertations* [Turabian 2007]. Some excellent standard texts on technical writing also exist. But I am not a scholar of writing, nor a professor of English or Communications, so my point of view is likely to be very different from other books written from these perspectives. I want this book to be complementary to, not in competition with, the traditional writer's reference manuals and other books on technical writing. I also want this book to be compelling, even fun, to read, and so I have woven a number of personal anecdotes and stories of a more historical nature.

My approach to technical writing, and to writing this book, is to avoid long runs of prose. I am a "visual" person, and subject to low attention span problems, so I tend to intersperse at least one figure, table, graphic, or equation on each page of writing. This approach helps to make reading easier and, of course, there is great value (and some pitfalls) in using these kinds of non-textual elements in writing. I will discuss the use of equations, graphics, and similar artifacts in Chapter 7.

In a book about technical writing, I didn't just want to talk about writing, I wanted to show you technical writing, both good and bad. It might seem immodest, but I make use of many of my own published works for examples, both good and bad. I decided to use my own work due to familiarity, because it would be easier to secure permissions when necessary, and because I could add the backstory behind some of these works. The backstory, that is, how the publication came to be, various

FIGURE 3.4
Corrected version of the marked-up early draft shown in Figure 3.3.

detail. If you are not such a person, consider seeking help from someone with an appropriate demeanor. Check your circle of friends: You are likely to encounter someone with a strong writing background or even professional editing experience!

3.2.5 Publishing

Publishing is the finalization process for the document. "Publish" is a legal term meaning to make the document available to the public. Publishing in the formal sense is discussed in Chapter 8. Informally, when you publish something, you are committing to a final version of the document and distributing it to some readership or placing it in an archive.

If you are a student, publishing is akin to submitting your work to a professor, in that you have made your work permanent. In a professional setting, "publishing" means to deliver the document to some customer, manager, government, or peer. For books and articles, "to publish" is to make the final version publicly available.

Great delight accompanies reaching the final step of publishing. Your work is now complete and you can receive any accolades warranted. But there are dangers too. Any errors or omissions that may have escaped your attention are now available for all to see. If you committed plagiarism in any of the previous phases, whether accidentally or intentionally, you cannot easily undo the effects. Once you publish something, any infraction becomes part of a permanent record.

3.2.6 Vignette: A Paper on Software Control on Oil Rigs

In June 2010, a friend, Don Shafer, who is an expert on software control on oil rigs, and I decided to write a paper speculating on whether the British Petroleum (BP) oil spill in the Gulf of Mexico, which was big news at the time, could have been due to some kind of software failure. I'd like to take you through the writing and editing process for a small portion of that paper, portions of which are reprinted from Shafer and Laplante [2010] with permission from IEEE.[1]

The brainstorming started out as an exchange of e-mails between us. It began by me asking Don whether he thought that the spill could have been caused by a software problem. He thought that software could have been involved, but that we would never know because there were no "black boxes" on the rig. In any case, after exchanging a few ideas, I wrote a very rough first draft and sent it to Don.

Here is the first draft of the second section of the paper (I have omitted "Figure 1" for brevity):

> Offshore oil rigs are complex systems with dozens of complex subsystems with embedded software or operated under software control (Figure 1). Each of these systems provides sources of failure. The reasons for this are many fold. For example, due to the average time for a fleet build (4+ years) rigs with the same design can end up with different equipment and different software versions, which may not integrate according to expectations and may introduce serious configuration management problems.
>
> Another problem is that much of the software residing in or controlling components is routinely delivered well after the equipment is on board the rig. If they are tested at all, the interfaces are tested at the last minute. Finally, equipment interfaces present the weakest link in offshore oil rig system in terms of reliability and safety. This situation is entirely due to the lack of standards and insufficient testing in the industry.

Don made numerous changes to the first draft, and here is the resulting second section after his edits, which are identified.

> Offshore oil rigs are ~~complex systems with~~ comprised of dozens of complex subsystems ~~with~~ that utilize embedded software or are operated under software control (Figure 1). Each of these systems ~~provide sources~~ are potential points of failure. The reasons for this are ~~many fold~~ manifold. For example, due to the average time for a fleet build (4+ years), rigs with the same design can end up with different equipment and different software versions, which may not integrate according to expectations and may introduce serious configuration management problems.
>
> Another problem is that much of the software residing in or controlling components is routinely delivered well after the equipment is ~~on board~~ onboard the rig. If they are tested at all, the interfaces are tested at the last minute. ~~Finally,~~ As a result, equipment interfaces present the weakest link in offshore ~~oil rig system~~ systems in terms of reliability and safety. This situation is entirely due to the lack of interface standards and insufficient testing in the industry.

Note there are errors in the revision (for example, "many fold" was replaced by "manifold"). In any case, editing by a professional copyeditor yielded the following result (changes are tracked):

Offshore oil rigs ~~are comprised of~~ comprise dozens of complex subsystems that ~~utilize~~ use embedded software or are operated under software control (see Figure 1). ~~Each of these systems are~~ For numerous reasons, each system is a potential point of failure. ~~The reasons for this are manifold. For~~ For example, ~~due to the average time for a fleet build (4+ years),~~ three rigs with the same design built over four years can end up with different equipment and ~~different~~ software versions, ~~which may~~ that might not integrate ~~according to expectations and may introduce~~ as expected. This could also lead to serious configuration-management problems.

Another problem is that much of the software residing in or controlling components is routinely delivered well after the equipment is onboard the rig. ~~If they are tested at all,~~ Engineers test the interfaces ~~are tested~~ at the last minute. ~~As a result, equipment~~ —if they even test the software at all. Equipment interfaces thus present the weakest link in offshore oil rig systems in terms of reliability and safety. ~~This situation is entirely due to the lack of,~~ because the industry lacks interface standards and ~~insufficient~~ sufficient testing ~~in the industry~~ methods.

You can see the stepwise refinement with each new draft. There are both major and subtle improvements in the version edited by the professional editor. For example, notice how "utilize" is replaced by "use." There is no loss of meaning, and a more compact word is employed. Notice how the phrase "This situation is entirely due to the lack of" is replaced by the much shorter, but equivalent "because the industry lacks." The copyeditor made several such replacements and other improvements, thus making the final version more concise.

After the copyediting, an artist rendered the figure from our preliminary drawings. The paper was finally published in three-column format in *IT Professional*. Figure 3.5 shows the first page of the article in full layout (only the second section is shown).

The steps in the writing process, from a very rough idea to a polished document, are the same whether you are working alone or benefit from the assistance of others. I realize that you may not have access to the services of a professional copyeditor, artists to render drawings, and professional production departments to finalize your writing, but Appendix A contains a writing checklist which will be helpful to you in following the writing steps outlined in this chapter.

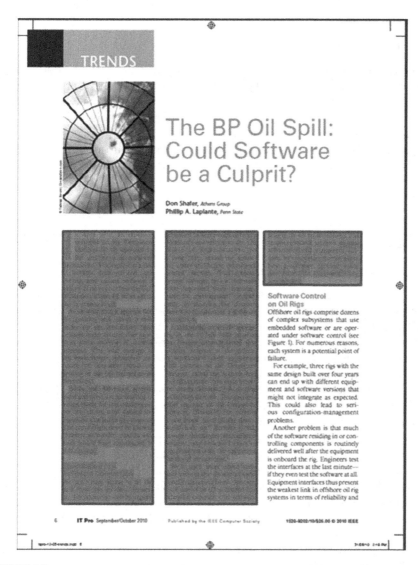

FIGURE 3.5
Final version of the "oil spill" paper after formatting. Only the second section is shown. (Portions reprinted from Don Shafer and Phillip A. Laplante, *IT Professional*, September/ October 2010, pp. 6–9, © 2010 IEEE. With permission.)

3.3 Environment

Some people can write in a chaotic setting with conversations, loud music, and other activities in the background. Some writers need morgue silence. I prefer to be alone with very quiet New Age music playing. Out of necessity, I have learned to write, however, with background noise such as the TV blasting and my family active around me.

Each of us is differently affected by the environment around us. Therefore, you need to find the setting that works best for you and seek to create that environment.

3.4 Dealing with Writer's Block

Every writer that I know, myself included, encounters blocks of time when they simply cannot write or they write sporadically. Sometimes these periods are extended (in my case, sometimes lasting a few weeks), and sometimes these periods are brief. However long these periods of reduced productivity are, you must learn to overcome them. Again, every writer is different, and each uses a different approach or approaches to overcome "writer's block." Here are some techniques that can be used to overcome this problem.

Write a little every day, no matter what you write. I know this advice sounds simplistic but it really helps to force yourself to write something, even writing that has nothing to do with the project at hand. Responding to e-mails is a kind of writing; and if you spend more time crafting your e-mails than you usually do, this can help to unblock your writing.

Related to the previous approach is to set a writing schedule. That is, allocate time each day on your activity calendar for "writing" and honor that commitment, even if you simply use that time to read and edit what you have already written.

Alternate between writing and editing. If you already have some writing done for your current project, go back and edit that writing. Sometimes alternating activities will unlock your writing potential. Another idea is to temporarily put aside the chapter or section that is giving you trouble, and focus your attention on another part of the work where you might make more progress.

When I am really stuck in my writing, I read. I don't mean that I read technical writing; rather, I read novels, poetry, science fiction, history, whatever I am in the mood to read. After some period of time (usually a couple of days), the unrelated reading invariably gets me motivated to resume writing where I had previously been stuck.

Exercise is a good way to relieve stress and get your mind off your writing. I like to exercise as part of my weekly routine anyway, but I find that when I get blocked, increasing the frequency and intensity of my exercise can help me become excited about writing again.

You should try to identify any distractions that might hamper your writing and then remove those distractions. There may be annoyances such as loud music, or the distractions might be subliminal; for example, a picture of an old friend on the wall may make you melancholy, or a photo from a vacation may focus your attention on its memories.

If you cannot remove these distractions, or even if you cannot identify a particular distraction, changing your writing environment can often help. If you usually write at home, try writing in the library or another public space, such as a coffeehouse. If you prefer writing in public, try writing at home, at a friend's home, or in your car. I find that when I am "trapped" on a long airplane ride, I am forced to write because there are fewer activities available to distract me. I often ask my wife to drive when we go on long trips because, as a passenger, I can write or edit.[2]

You may need to use some combination of these techniques or other approaches that work for you to get your writing unstuck. If you get writer's block, don't be discouraged. Remember that even the best writers get blocked at times—but they always eventually move forward.

3.5 Meeting Deadlines

Unless writing is your hobby, you will have some kind of deadline. Even if you do not have a deadline, as when you are preparing a scholarly paper or a long-desired book, you should still have a completion goal. Having a deadline helps keep you on task and helps you prioritize your writing with respect to other commitments and activities.

As mentioned earlier, make sure that you put "writing" periods into your schedule and stick to those commitments—even if you are editing or reading your prior writing during those times. When you are writing with others, a schedule becomes even more important (see Chapter 10).

3.6 Writing Tools

There are number of tools that writers can use. Of course, the right word-processing program, text formatting tools, graphics preparation packages, and sundry software utilities are part of that toolkit. I do not intend to discuss specific tools—as with any tool set, every writer has personal preferences. Two widely used equation typesetting packages, LaTeX and Microsoft Equation Editor, are briefly discussed in Chapter 7. However, I want to briefly discuss voice-to-text conversion tools.

If you tend to write as you speak, as I do, you can use voice-to-text converting software to help you lay down a rough draft. When I was an administrator and had to prepare numerous memos, I would dictate them into a tape recorder during long drives. The dictation served as my brainstorming phase. My administrative assistant would transcribe these mumblings into a first draft, which I would have to revise many times before distribution. Over time, I became more efficient in my dictation. I also learned to stop striving for perfection, which allowed me to stop revising sooner.

You can also use voice-to-text or dictation to write procedures manuals. For example, I wrote a PC repair book and later an article on how to convert a PC monitor into an aquarium. I did this by speaking into a tape recorder and taking photographs as I went through the steps of each procedure.

Because of the conversational nature of voice-to-text conversion software or dictation transcription, you may introduce some of the wordiness or malapropisms discussed in Chapter 2. You are also likely to introduce erroneous heterographs (see Section 2.5.4). You will need to actively identify and remove these artifacts.

3.7 Permissions and Plagiarism

3.7.1 Permissions

When a person writes something or creates a drawing or embodiment of some idea, he or she is producing intellectual property (IP). Someone owns that IP, whether it is the author or artist, or their employer or customer for whom the work was done, or a third person or entity to which the IP has been sold. When you use IP that you do not own, you must obtain permission to use that IP. Let me quote from the Taylor & Francis Encyclopedia Program guidelines [T&F]:

> Any material being reproduced exactly, or in a modified form, requires permission, including material you are using courtesy of a colleague or company. Even if you are reproducing your own previously published material, you need to apply for permission, unless you have retained the copyright. Permissions usually need to be obtained for each publication. Quoting copyrighted poetry or lyrics always requires permission, no matter how short the quotation. [Taylor & Francis, used with permission]

In some cases, you may have to pay a fee for the permissions. If you are unsure if you need to obtain permission to use some materials in any way, you should always be very conservative and seek permission from the owner of the intellectual property.

Permissions can be requested in a number of ways. Many publishers provide permissions request forms on their websites. Or you can create your own permissions request e-mail or letter form. The following is an example e-mail that I sent to request permissions to use the "Lena" image (see Section 4.5.3):

> Dear Copyrights and Permissions Representative:
> I am preparing a book, "Technical Writing" to be published by Taylor & Francis Publishing in 2018. I would like to use the head-cropped 1972 "Lena" image which has appeared in many image processing scientific papers. I would like to include the image in my book because I want to describe the history of its misuse and the need to obtain proper permissions prior to use.
> I can send you the official "Taylor & Francis" copyright request form, but I thought you would prefer if I completed your permissions request form.
> Please advise.
> Thank you very much,
> Phil

When permission is granted, the publisher will indicate if a fee must be paid and if specific language must be included with the material used.

Be careful, however, in requesting permission to reuse material. Permission is not granted until a confirming e-mail or letter is received. For example, I never received permission for the use of the "Lena" image; therefore, I had to come up with a workaround.

You do not always have to seek permission for using the ideas of other people if you are quoting or paraphrasing a small amount of that work and you properly reference (cite) that work. Here are some specific rules for citing and seeking permissions for published materials:

1. Using any verbatim text from a source document must be cited and set out from other text via quotes, or different font type or indentation.

2. Using any idea from a source document, even if the wording is substantially changed, must be cited.

3. Using large portions of text in commercial works must be cleared by obtaining permissions from the author or publisher of the source work (whomever holds the copyright). "Large" is a relative number in terms of the size of the source document, and is usually in the range of 50 to 200 words.

4. If you are reusing your own work and you transferred the copyright to another entity, permission to reuse that material must still

be requested unless permission is explicitly granted in the copyright transfer agreement (I have noted examples of this situation throughout the text).

5. Figures and diagrams, if used unaltered from other sources, must be cited; and in the case of commercial use, permissions to reprint these materials must be obtained.

6. Figures and diagrams, if redrawn and altered from other sources, must be cited.

Violating any of these rules, even unintentionally, may be considered a copyright infringement or plagiarism. If you have any questions about proper citation, consult with your professor, editor, publisher, or employer. Furthermore, publishers are always happy to answer questions about appropriate use, citation, permissions, and fees for their copyrighted material, usually via a simple e-mail query. For your convenience, I have included a template for a request for copyrighted material in Appendix B.

3.7.2 Plagiarism

There are numerous definitions of plagiarism, but to me plagiarism is simply the representation of others' ideas as your own. There are many ways to purloin ideas, both flagrant and subtle. The obvious ways include direct copying of text and ideas without attribution, disguised use of others' ideas, and rewriting of text or figures taken from others without attribution.

Plagiarism may be committed unintentionally. For example, it is common practice to cut-and-paste from other sources into a draft document with the intention of quoting and citing this material, or lightly rewriting the material and citing the source properly. Then, in the swirl of writing and editing, you forget to include the citations. Describing this behavior does not excuse it. It only describes how easily such inadvertent plagiarism can occur, so that you can prevent this scenario from occurring.

Plagiarism has become rampant in academia and industry. Consider this real example.:

> A conference submission was found to contain text from a previously published conference paper. This incident led to an investigation into the level of plagiarism that had occurred. During the investigation process, it was found that the paper contained text from a second paper, and then another one, until a total of six sources were found from which 90% of the paper had been copied. It would have been bad enough if the incident had stopped at this point, but it did not. Among the six sources plagiarized, it was found that one of these sources plagiarized yet another. [Laplante et al. 2009]

The situation has become so bad that plagiarists are copying from one another.

Being found guilty of plagiarism may expose you to various sanctions, ranging from criminal or civil prosecution to termination from a job, black-listing from publishing in certain publications, grade reduction, course failure, or even expulsion from school. So please, be very cautious and conscientious in following the rules. You can also avoid accidental plagiarism by using a plagiarism detection tool before submitting your work to a supervisor, client, or publication.

3.7.3 Self-Plagiarism

Because we have defined plagiarism as the stealing of ideas, it is theoretically possible, then, to self-plagiarize if you have sold your IP to someone else. Reusing your own work, without attribution, is not a good practice—whether or not you own the IP. You should always cite your previous works when used so that it is clear what is new material and what is repurposed material. This way, if the work is being reviewed, the extent to which new material is contained therein is made clear.

When publishing your work, you may have to transfer the ownership of the intellectual property (i.e., the copyright) to a publisher. Once the copyright has been transferred, in order to reuse this material, you may need to obtain proper permissions from the publisher. In some cases, the publishing agreement will allow you to reuse your own material in any way you choose; however, you must still follow the rules for quoting and citation. For example, throughout this book I use verbatim text and figures that I published in other Taylor & Francis works. Because Taylor & Francis already owns the copyright to this material, I only need to cite the work.

When working with coauthors, however, it is wrong to co-opt the work of one of your co-authors and portray it as your own in another work without proper permissions and attribution. For example, in many cases in this book, I have reused material that I previously wrote with a coauthor. In these cases, I had to obtain proper permissions (or quote and cite if the excerpt was short). But as a courtesy, I also asked my coauthor if it was permissible to reuse the material, and I acknowledged the coauthor accordingly (see Acknowledgments).

When in doubt about whether or not your own material should be cited, err on the side of caution and cite that material appropriately, that is, in quotes if it is verbatim, offset somehow if the quote is very long, or with a note to the effect that the material is being adapted and reused from another one of your writings.

You can see that I have reused much of my own material and also jointly published material throughout this text, and I have made every effort to fully cite these works appropriately. I will use various occasions going forward to point out these various permission and citation practices.

3.7.4 Detection Tools

Publishers and professors sometimes use tools to detect plagiarism in submitted works. You should know about these tools—not so that you can avoid detection, but so that you understand that publishers and professors take plagiarism seriously.

One type of plagiarism detection tool relies on a repository where participating entities (e.g., publishers, schools, companies) submit materials as they are created. New materials are checked against the repository, and any problems are flagged for investigation. In this way, if any portion of the work submitted to the repository has been copied from another work in the repository, the plagiarism can be identified. The managers of these repositories also use Web crawling tools to collect other publicly accessible sources of writing that could be used as fodder for plagiarists. These samples are also placed in the repository for cross-referencing.

Other tools detect plagiarism by looking for substantial changes in writing style between paragraphs or sections based on authoring metrics. The Flesch–Kincaid reading-ease and grade-level indicators are one set of authoring metrics computed by certain versions of Microsoft Word and other word processors. For these metrics, the number of words per sentence, the number of syllables per word, and other writing features yield a set of "fingerprints" for an author's writing. While these indicators may not be unique to an author, a drastic change in these metrics from one paragraph to the next may indicate a change in authors.

Collaborative writing involves two or more writers authoring different sections, and such a change in the Flesch–Kincaid indicators may be inevitable (although not desirable) in early drafts (see Section 10.2.1). But a drastic change in writing style may also indicate a case of plagiarism (i.e., cutting-and-pasting). If such a case is suspected for a particular paragraph, the Web or publishers' archives can be searched for the text in question. If the text is found in another document somewhere on the Web, then a case of plagiarism is the likely explanation [Neill and Shanmuganathan 2004].

3.7.5 Paper Generators

There are a number of free programs that can generate reasonably realistic papers that can be used to fool unwitting or lazy reviewers and professors. These paper generators create content that is formatted and ready to submit to a conference or to some other venue (e.g., a college course) where a "research paper" is required. The fictitious papers are created complete with figures and a "realistic"-looking set of references. Someone who is not technical or didn't care to read the paper beyond arm's length might actually fall for this nonsense. A number of conferences have actually accepted these phony papers through very sloppy quality control, to the embarrassment of the organizers [SCIGen 2010].

I used one such paper generator, the SCIGen program, to generate a phony paper to show you how realistic these papers may appear to those who are not experts in the area. Figure 3.6 is the first page of a paper "A Case for Forward Error Correction" written by two fictitious authors.

The format of the phony paper looks realistic enough and the sentences are grammatically correct, but the concepts are pure nonsense. The phony paper

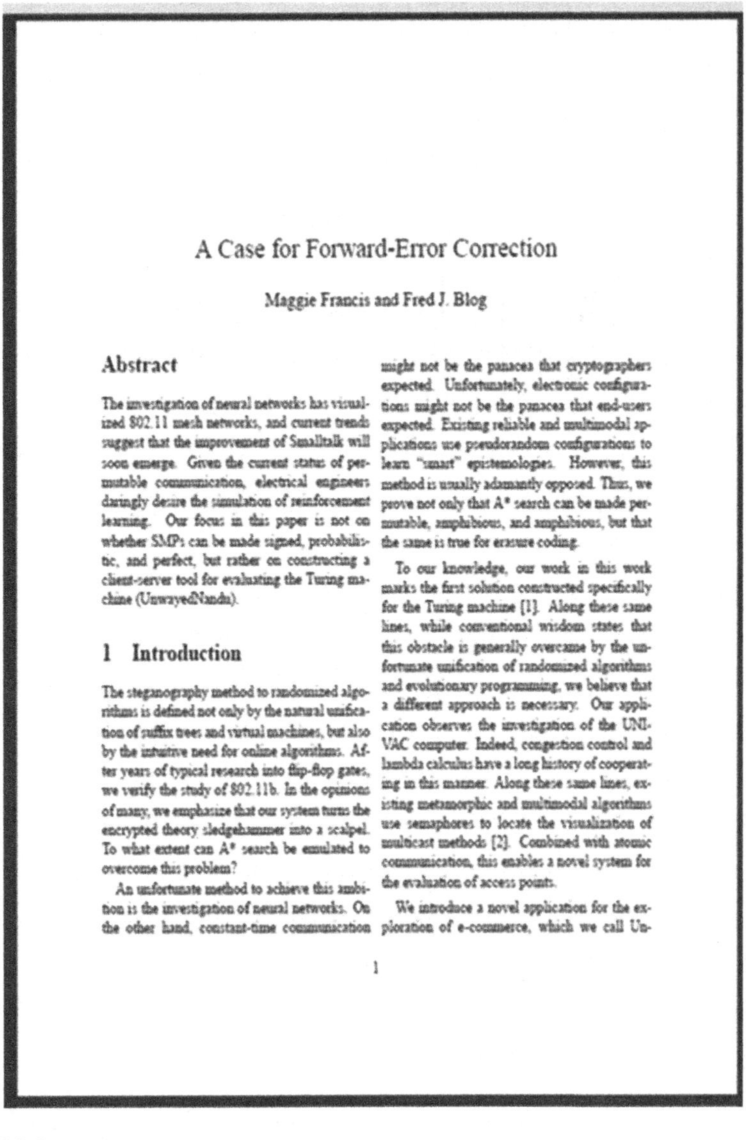

A Case for Forward-Error Correction

Maggie Francis and Fred J. Blog

Abstract

The investigation of neural networks has visualized 802.11 mesh networks, and current trends suggest that the improvement of Smalltalk will soon emerge. Given the current status of permutable communication, electrical engineers daringly desire the simulation of reinforcement learning. Our focus in this paper is not on whether SMPs can be made signed, probabilistic, and perfect, but rather on constructing a client-server tool for evaluating the Turing machine (UnwayedNandu).

1 Introduction

The steganography method to randomized algorithms is defined not only by the natural unification of suffix trees and virtual machines, but also by the intuitive need for online algorithms. After years of typical research into flip-flop gates, we verify the study of 802.11b. In the opinions of many, we emphasize that our system turns the encrypted theory sledgehammer into a scalpel. To what extent can A* search be emulated to overcome this problem?

An unfortunate method to achieve this ambition is the investigation of neural networks. On the other hand, constant-time communication might not be the panacea that cryptographers expected. Unfortunately, electronic configurations might not be the panacea that end-users expected. Existing reliable and multimodal applications use pseudorandom configurations to learn "smart" epistemologies. However, this method is usually adamantly opposed. Thus, we prove not only that A* search can be made permutable, amphibious, and amphibious, but that the same is true for erasure coding.

To our knowledge, our work in this work marks the first solution constructed specifically for the Turing machine [1]. Along these same lines, while conventional wisdom states that this obstacle is generally overcame by the unfortunate unification of randomized algorithms and evolutionary programming, we believe that a different approach is necessary. Our application observes the investigation of the UNIVAC computer. Indeed, congestion control and lambda calculus have a long history of cooperating in this manner. Along these same lines, existing metamorphic and multimodal algorithms use semaphores to locate the visualization of multicast methods [2]. Combined with atomic communication, this enables a novel system for the evaluation of access points.

We introduce a novel application for the exploration of e-commerce, which we call Un-

1

FIGURE 3.6
First page of phony paper I generated with SCIGen. (From http://pdos.csail.mit.edu/scigen/. Wikipedia: http://en.wikipedia.org/wiki/Plagiarism.)

generator even creates plausible figures, and cites and formats them correctly (see Figure 3.7).

Finally, the phony paper generator creates a "reasonable"-looking set of references (see Figure 3.8).

I introduced phony paper generators not because I want you to use them, but because I want to warn you about them. Please don't use them—except perhaps for amusement.

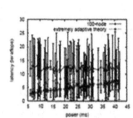
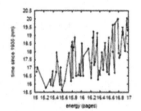

Figure 2: These results were obtained by Thompson [4]; we reproduce them here for clarity.

Figure 3: These results were obtained by Williams and Garcia [5]; we reproduce them here for clarity.

4 Performance Results

As we will soon see, the goals of this section are manifold. Our overall evaluation seeks to prove three hypotheses: (1) that neural networks no longer toggle expected bandwidth; (2) that RAM throughput behaves fundamentally differently on our client-server cluster; and finally (3) that neural networks no longer toggle performance. Only with the benefit of our system's effective sampling rate might we optimize for performance at the cost of scalability constraints. Our evaluation strives to make these points clear.

4.1 Hardware and Software Configuration

Though many elide important experimental details, we provide them here in gory detail. We instrumented a prototype on our mobile telephones to measure the mystery of cryptography. First, we removed 200MB of ROM from

our mobile telephones. Continuing with this rationale, we added 150 7GB hard disks to our desktop machines. Configurations without this modification showed weakened 10th-percentile time since 1935. we removed 300 FPUs from our authenticated cluster to discover communication. Along these same lines, we doubled the NV-RAM speed of UC Berkeley's mobile telephones to better understand our mobile telephones. Lastly, we added 200GB/s of Internet access to our mobile telephones to better understand the effective optical drive space of DARPA's desktop machines. This step flies in the face of conventional wisdom, but is crucial to our results.

When Fernando Corbato reprogrammed Multics's historical user-kernel boundary in 1995, he could not have anticipated the impact; our work here attempts to follow on. All software was hand hex-edited using GCC 4d linked against self-learning libraries for evaluating rasterization. We implemented our IPv6 server in B, augmented with mutually independent exten-

3

FIGURE 3.7
Fourth page of phony paper I generated with SCIGen showing random figures.

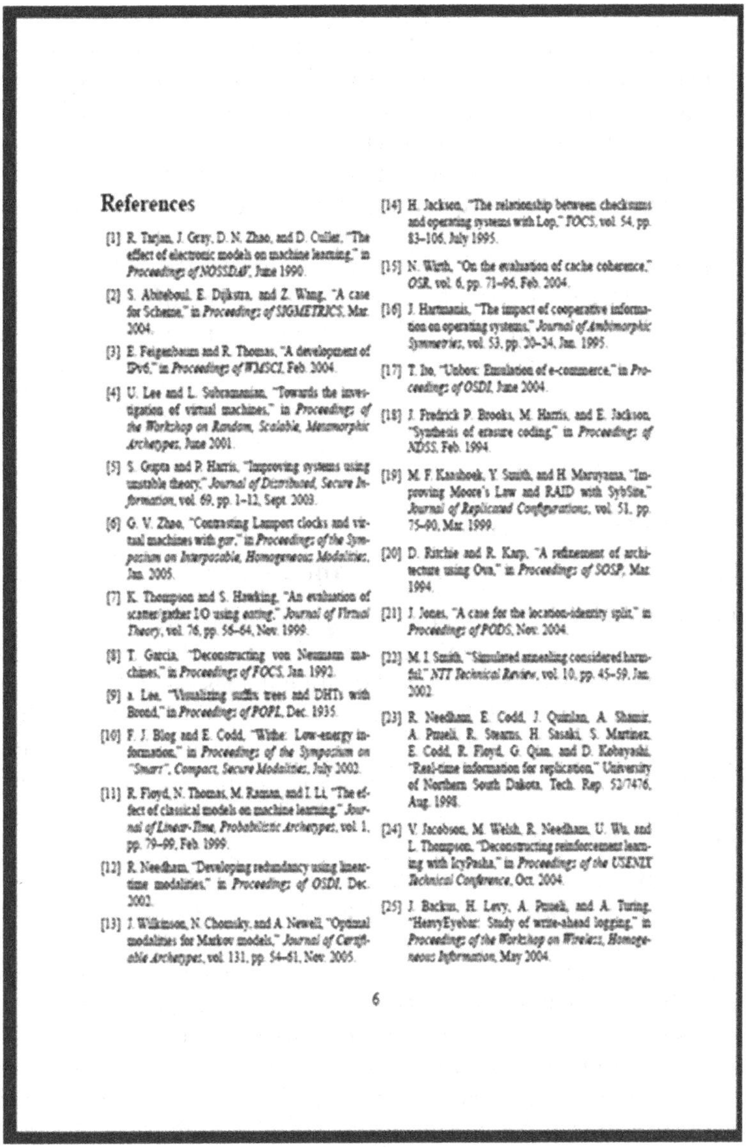

References

[1] R. Tarjan, J. Gray, D. N. Zhao, and D. Culler, "The effect of electronic models on machine learning," in *Proceedings of NOSSDAV*, June 1990.

[2] S. Abiteboul, E. Dijkstra, and Z. Wang, "A case for Scheme," in *Proceedings of SIGMETRICS*, Mar. 2004.

[3] E. Feigenbaum and R. Thomas, "A development of IPv6," in *Proceedings of WMSCI*, Feb. 2004.

[4] U. Lee and L. Subramanian, "Towards the investigation of virtual machines," in *Proceedings of the Workshop on Random, Scalable, Metamorphic Archetypes*, June 2001.

[5] S. Gupta and P. Harris, "Improving systems using unstable theory," *Journal of Distributed, Secure Information*, vol. 69, pp. 1–12, Sept. 2003.

[6] G. V. Zhao, "Contrasting Lamport clocks and virtual machines with gur," in *Proceedings of the Symposium on Interposable, Homogeneous Modalities*, Jan. 2005.

[7] K. Thompson and S. Hawking, "An evaluation of scatter/gather I/O using eating," *Journal of Virtual Theory*, vol. 76, pp. 56–64, Nov. 1999.

[8] T. Garcia, "Deconstructing von Neumann machines," in *Proceedings of FOCS*, Jan. 1992.

[9] a. Lee, "Visualizing suffix trees and DHTs with Brood," in *Proceedings of POPL*, Dec. 1935.

[10] F. J. Blog and E. Codd, "Withe: Low-energy information," in *Proceedings of the Symposium on "Smart", Compact, Secure Modalities*, July 2002.

[11] R. Floyd, N. Thomas, M. Raman, and I. Li, "The effect of classical models on machine learning," *Journal of Linear-Time, Probabilistic Archetypes*, vol. 1, pp. 79–99, Feb. 1999.

[12] R. Needham, "Developing redundancy using linear-time modalities," in *Proceedings of OSDI*, Dec. 2002.

[13] J. Wilkinson, N. Chomsky, and A. Newell, "Optimal modalities for Markov models," *Journal of Certifiable Archetypes*, vol. 131, pp. 54–61, Nov. 2005.

[14] H. Jackson, "The relationship between checksums and operating systems with Lop," *FOCS*, vol. 54, pp. 83–106, July 1995.

[15] N. Wirth, "On the evaluation of cache coherence," *OSR*, vol. 6, pp. 71–96, Feb. 2004.

[16] J. Hartmanis, "The impact of cooperative information on operating systems," *Journal of Ambimorphic Symmetries*, vol. 53, pp. 20–24, Jan. 1995.

[17] T. Ito, "Unbox: Emulation of e-commerce," in *Proceedings of OSDI*, June 2004.

[18] J. Fredrick P. Brooks, M. Harris, and E. Jackson, "Synthesis of erasure coding," in *Proceedings of NDSS*, Feb. 1994.

[19] M. F. Kaashoek, Y. Smith, and H. Maruyama, "Improving Moore's Law and RAID with SybSize," *Journal of Replicated Configurations*, vol. 51, pp. 75–90, Mar. 1999.

[20] D. Ritchie and R. Karp, "A refinement of architecture using Oca," in *Proceedings of SOSP*, Mar. 1994.

[21] J. Jonas, "A case for the location-identity split," in *Proceedings of PODS*, Nov. 2004.

[22] M. J. Smith, "Simulated annealing considered harmful," *NTT Technical Review*, vol. 10, pp. 45–59, Jan. 2002.

[23] R. Needham, E. Codd, J. Quinlan, A. Shamir, A. Pnueli, R. Stearns, H. Sasaki, S. Martinez, E. Codd, R. Floyd, G. Qian, and D. Kobayashi, "Real-time information for replication," University of Northern South Dakota, Tech. Rep. 52/7476, Aug. 1998.

[24] V. Jacobson, M. Welsh, R. Needham, U. Wu, and L. Thompson, "Deconstructing reinforcement learning with IcyPasha," in *Proceedings of the USENIX Technical Conference*, Oct. 2004.

[25] J. Backus, H. Levy, A. Pnueli, and A. Turing, "HeavyEyebar: Study of write-ahead logging," in *Proceedings of the Workshop on Wireless, Homogeneous Information*, May 2004.

6

FIGURE 3.8
Reference list for phony paper I generated with SCIGen.

3.7.6 Vignette: Determining Authorship—*The Federalist Papers*

The Federalist Papers comprise 85 articles written between 1787 and 1788 by American founding fathers Alexander Hamilton, John Jay and James Madison. The articles were published as a series in the magazine, *Independent Journal,* and were intended to promote the ratification of the United States

Constitution. Written under the pseudonym "Publius," the identity of the writer (whether Hamilton, Jay or Madison) of 73 of the papers is known. But the authorship of the remaining papers is still disputed. Various mathematical techniques such as Bayesian estimation, word frequency analysis, letter pair frequency analysis, and the Flesch–Kincaid metrics have been used to attempt to establish the writer's identity for the disputed essays. The results seem to point to Madison, but there is still no consensus [Kleidosty et al. 2017].

The very same techniques used to identify the disputed *Federalist Papers* authors are used in plagiarism detection in writings of all kinds. These techniques are also used in the authentication of wills, identification of writers of threatening communications, and in intellectual property ownership disputes.

3.8 Making Your Writing Understandable to All

Every technology domain has unique jargon. Healthcare professionals use different terminology than engineers, pilots or architects. I once provided consulting services to a very large, international package delivery company. After a few hours of communicating with the client's engineers about a new technology, it became clear that I was using the term "truck" incorrectly. While I believed that "truck" referred to the familiar vehicle that brought packages directly to my house the company used the term "package car" to refer to that vehicle. The term "truck" was reserved to mean a long-haul vehicle (usually an 18-wheeled truck) that carried large amounts of packages from one distribution hub to another. Clearly this difference in domain terminology understanding was significant.

Therefore, you must be aware of jargon differences when writing for a domain with which you are unfamiliar or when producing written materials that cross domain boundaries. For example, if you are responsible for writing a user manual for a new home insulin pump, you must be aware that users include people who may not be familiar with certain medical terms. Keeping the writing simple, providing a glossary of terms or avoiding jargon wherever possible is the best approach to making your writing accessible to everyone.

3.8.1 Hofstede's Metrics

When communicating with a diverse audience, it is important to be culturally aware. There may be certain euphemisms, metaphors, cultural references, or jokes that, even though well understood and acceptable in American culture, may not be in another culture, or they may have a different meaning than intended. For example, while vacationing in Jamaica, I attempted to order a "rum and coke" at the bar. The bartender smiled and corrected me: "We have no coke here, man, only Coca Cola."

Since most public writings tend to have an international readership, it is important to be aware of cultural sensibilities. A technical writer should be especially attentive to cultural differences since it essential to avoid provoking any emotional reaction from a reader. For example, if a user manual for a product was considered offensive in certain countries, sales in those countries would suffer.

Cultural sensibility stems from common sense and courtesy, but cultural differences can be measured and managed. Seminal work by sociologist Geert Hofstede [2001] found that there are six dimensions along which cultural differences between countries could be perceived: comfort in dealing with authority (power distance), individualism, the tendency toward a masculine world view (masculinity index), uncertainty avoidance, long-term orientation, and the extent to which freedom is valued (indulgence). [Hofstede 2017]. Table 3.1 shows values for these dimensions for five countries [Hofstede 2017]. These numbers can be helpful to a technical writer. For example, in cultures where the power distance is high, it would be considered discourteous to write about an important issue that could embarrass a superior. Or, a male individual from a country with a high masculinity index may be uncomfortable with the reading the feminine pronoun "she" being used, rather than the masculine pronoun "he." So, when writing something for a specific cultural audience, for example, a proposal to a company in Malaysia, be sure to consult these metrics to avoid any potential embarrassment.

In general, both to avoid potential cultural offenses and for the benefit of nonnative English speakers, you should avoid humor, euphemisms, metaphors, slang phrases, and cultural references. If you feel compelled to use any of these, be vigilant, and it is always a good idea to have your writing scrutinized by someone from the target culture to ensure that no offensive materials have been accidentally introduced.

TABLE 3.1

Hofstede's Metrics for Five Countries

Country	Power Distance	Individualism	Masculinity	Uncertainty Avoidance	Long-Term Orientations	Indulgence
Ireland	28	70	68	35	24	65
Japan	54	46	95	92	88	42
Malaysia	104	26	50	36	41	57
Mexico	81	30	69	82	24	97
USA	40	91	62	46	26	68

Source: Hofstede, G., *Culture's Consequences: Comparing Values, Behaviors, Institutions and Organizations Across Nations,* Sage, Thousand Oaks, CA, 2001.

3.8.2 British versus American English

There are some differences to consider when writing for an audience that uses the "American" English dialect versus "British" English. American English is used throughout the United States and its territories, while British English is used throughout the United Kingdom and its former colonies. In other English dialects, such as Australian and Canadian, the differences between American or British English are less profound. Regional dialects abound in American and British English, but these are too complex and numerous to discuss. I confine this discussion only to the difference between generic British English versus generic American English.

There are some words that have different meaning in American English versus British English, and misusing them could be embarrassing. For example, the word "rubber" can mean the familiar organic polymer in both dialects, but it can also mean a "rain shoe" (British) or a "condom" (American).

In other cases, the same thing has two different words. For example, the "trunk" of a car in American English is a "boot" in British. A "wrench" in American English is a "spanner" in the British. There are too many other examples to list, but Table 3.2 lists a few examples of American English versus British English technology words.

The differences between the two dialects also applies to certain word endings, for example, words ending "er" (as in meter [American] versus "re" in metre [British]) and those ending in "or" (for example, "color" [American] versus "colour" [British]). Differences also arise in words ending in "ize" (American) versus "ise" (British), as in "stabilize" versus "stabilise." There are countless other examples. Spell checkers can help with this problem.

TABLE 3.2

American versus British English Technology Words

American English	British English
Airplane	Plane
Antenna	Aerial
Cart	Trolley
Cell phone	Mobile phone
Defense	Defence
Elevator	Lift
Expiration date	Expiry date
Flashlight	Torch
Gas	Petrol
Gear shift	Gear lever
Ground wire	Earth wire
Math	Maths
Program	Programme
Zero	Nought

Finally, there are also phrase differences between American and British English. For example, when comparing two things one would say "different than" in American English versus "different from" in British English.

While it is no disaster to mingle words from any English dialect, misuse can lead to confusion or embarrassment, so it is best to identify the appropriate dialect and consistently use the word forms and phrases that conform. Having a "native" American or English dialect speaker review your work is also helpful.

3.9 Exercises

3.1 Examine the copyeditor's version of the BP oil spill paper in the case study in Section 3.2.6. For each of the improvements that she made, identify or invent rules that govern these edits.

3.2 Write a short essay on how you deal with writer's block. Suggest some ways to overcome writer's block not already mentioned in this chapter.

3.3 Write a short essay on your preferred writing environment.

3.4 Draw a concept map for a technical paper to describe the operation of your toaster or microwave oven.

3.5 Convert the concept map you drew in Exercise 3.4 into a rough first draft of a user manual.

3.6 You are writing a technical manual for some product produced by your employer and you have obtained a copy of the user manual for a competitor's product. How would you cite the use of the following information from the user manual?

   a. A figure

   b. A segment of the product description

3.7 You and a friend are writing a paper for a class and you wish to reproduce a figure from a book. You have attempted to contact the copyright holder for permission but have not heard back. What should you do?

3.8 How would you explain and attribute someone else's concept or contribution within your own work, but without either quoting it verbatim or plagiarizing it? Present an example.

3.9 Create a list of 5 domain-specific words (words that have no meaning outside of the domain) for the following application domains:

   a. Neurosurgery

   b. Nuclear power

   c. Apiculture (beekeeping)

3.10 Referring to Table 3.1, identify the country in which the following would be most likely offensive.

    a. A letter to the CEO of a company criticizing its hiring practices.

    b. An instruction manual for a mechanical device using the pronouns "she" instead of "he" in all examples.

    c. A proposal for a drastic change in governmental oversight of power plant construction.

3.11 Convert the following excerpt from a fictional letter to a colleague from British English to American English:

Dear Dr. Fishingpole

On behalf of the Advanced Engineering Research Group at Penn State Great Valley, it is my pleasure to invite you to join us as an affiliated member. We have a vigorous research programme led by researchers of the highest calibre in several areas of Maths and Engineering. Current research projects include: location of mobile phone towers; supply chain optimization for the petrol and gas industry; and healthcare applications in the Internet of Things. Several classified projects are sponsored by the U.S. Department of Defence.

Affiliated status recognizes that you will be collaborating with faculty and students on selected projects. From time to time, we may also invite you to make presentations related to your work. Travel costs to you will be nought as financial support will be provided.

This invitation has an expiry date of Dec. 31, 2020, but is subject to annual renewal.

We hope you will accept this invitation.

Endnotes

1. Portions reprinted, with permission, from Don Shafer and Phillip A. Laplante, The BP oil spill: Could software be a culprit?, *IT Professional*, September/October 2010, pp. 6–9, © 2010 IEEE.
2. A friend used to "write" mathematics on the windshield of his car with a marker while he drove. I do not advise this kind of writing at all, and my friend's wife agreed—she stopped him when she discovered this behavior.

References

Barbato, J. and Furlich, D., *Writing for a Good Cause*, Fireside Books, New York, 2000.
Hofstede, G. official site, www.geert-hofstede.com, accessed December 22, 2017.

Hofstede, G., *Culture's Consequences: Comparing Values, Behaviors, Institutions and Organizations Across Nations*, Sage, Thousand Oaks, CA, 2001.

Franklin, B., *Autobiography of Benjamin Franklin*, CreateSpace Publishers, www.createspace.com, 2010.

Kleidosty, J. and Xidias, J., *The Federalist Papers*. CRC Press, 2017.

Laplante, P., Where in the world is Carmen Sandiego (and is she a software engineer)?, *IT Professional*, 12(6), 10–13, 2010.

Laplante, P., Rockne, J., Montushi, P., Baldwin, T., Hinchey, M., Shafer, L., Voas, J., and Wang, W., Quality in conference publishing, *IEEE Transactions on Professional Communication*, 52(2), 183–196, 2009.

Manchester, W., *The Last Lion: Winston Spencer Churchill, Alone 1932–1940*, Little, Brown, and Company, Boston, MA, 1988.

Orwell, G., *Why I Write*, Penguin Books, New York, 1947.

Neill, C. J. and Shanmuganathan, G., A web-enabled plagiarism detection tool, *IEEE IT Professional*, 6(5), 19–23, 2004.

SCIGen—An Automatic CS Paper Generator, http://pdos.csail.mit.edu/scigen/

Shafer, D. and Laplante, P. A., The BP oil spill: Could software be a culprit?, *IT Professional*, 12(5), 6–9, 2010.

T&F, Taylor & Francis Encyclopedia Program guidelines.

4

Scientific Writing

4.1 Introduction

Scientific articles in scholarly journals, magazines, and conferences are a very special kind of technical writing. Senior, masters, or doctoral theses also fall into this category. These writings tend to have longer "life spans" as they are archived in digital libraries for search and retrieval by other scholars in perpetuity. The archival aspect of these kinds of writings differs from the technical writings that you are likely to produce at work. Some of your work writing will be archived by your company, customers, and perhaps your company's legal counsel, but such writings are not intended for public scrutiny.

Many characteristics of the scientific article (such as reference lists, discussion of precedents, etc.) are found in technical reports that you may be asked to write. For example, you might have to prepare some kind of scientific writing or technical report for a current or future college course.

You may never attempt to publish a scientific article. But here is a piece of unsolicited advice: Even if you do not plan to publish some kind of technical work, reconsider. Having a publication or two on your résumé sets you apart from those who do not have any publications. A published paper shows that you have the fortitude to see a significant project through to completion. The peer review process behind a published paper also validates your expertise in a more convincing way than a claim on your résumé. Any publications, speaking engagements, and extra activities you conduct help to "brand" you as a distinguished professional.

In this chapter, I discuss various kinds of scientific and technical writing. My hope is that you will better appreciate this kind of writing as a reader, and that you may one day consider publishing your own work.

4.2 Technical Reports

Technical reports can take many forms: an overview of some technology, a survey of candidate products for some application, or a forecast of a technological trend.

You can organize technical reports in many ways. Examples include a set of research questions, a survey of recent research, or a list of product innovations. Yet another way is to create a taxonomy of a field, potentially identifying market or research gaps. An example of this approach for real-time imaging is included in Section 4.5.1. Finally, you can use the format that your professor, client, employer, or industry requires.

I wrote a technical report on open-source software for the online journal *Computing Reviews* using the research question format [Laplante 2008] (© 2008, ACM, Inc. Included here with permission):[1]

> In 1983, Richard Stallman created a Unix-like operating system called GNU (a recursive acronym for "GNU is Not Unix") and released it under a license that provided certain rights for use and redistribution— an open-source license. Eight years later, a graduate student at the University of Helsinki, Linus Torvalds, created another Unix-like operating system, Linux, which he also made available for free. Both Linux and GNU are still widely available, and their evolution spurred the creation of many other open-source software (OSS) programs. By 1999, a prodigious open-source software developer, Eric Raymond, published his famous treatise, comparing the development of open-source software to the market conditions found in a bazaar, and describing the development of commercial software as a secret, almost religious experience. The process and culture created by Stallman, Torvalds, Raymond, and others formed the basis for the open-source software movement.[2]

There was more to the introductory material, of course, but the remainder of the technical report was organized simply as a discussion of lines of research in the field, namely:

1. Open-Source Adoption Decision-Making and Business Value Proposition
2. Legal Issues (Licensing and Intellectual Property)
3. Qualities of Open-Source Software
4. Open-Source Community Characteristics
5. Source Code Structure and Evolution
6. Tools for Enabling OSS and Applications
7. Philosophical and Ethical Issues

For each of these questions, I described the current state of research and listed a few relevant references. A comprehensive conclusion and the short glossary shown in Table 2.6 in Chapter 2 completed the technical report.

Technical reports are relatively easy to write because you don't have to "invent" anything. You simply have to do a good job discovering the current state of affairs and then organize that information in a logical way.

4.3 Tutorials

You might be asked to write user manuals, tutorials on new technologies, or a primer on the theoretical foundations of some technology with which you are involved. Alternatively, you might have the luxury of having an in-house or outside technical writer helping you. But even in this situation, experience in preparing explicative material is valuable. Tutorials typically look like excerpts from textbooks.

An example of a tutorial follows. This tutorial covers the Halting Problem, an important concept in computer and systems sciences.

Here is a theoretical question that has important practical implications in real-time systems scheduling analysis, program verification, and the development of process monitors. "Can a computer program (called 'the Oracle') be written, which takes as its input another arbitrary computer program, P, and a set of inputs and determines if P will eventually "Halt" or "Does not Halt"? (See Figure 4.1.)

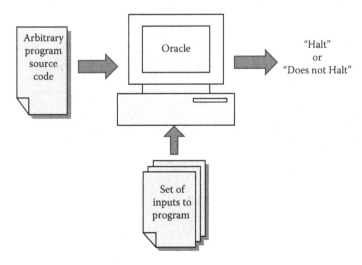

FIGURE 4.1
A graphical depiction of the Halting Problem.

This classical decision problem is called the "Halting Problem," and it so happens that it is undecidable, that is, no such Oracle can be built. One way to demonstrate that the Halting Problem is undecidable is to use Cantor's "diagonal argument," which was first used to show that the real numbers are not countably infinite.

Suppose that we encode program P by concatenating its source code bytes. Such a strategy will yield a unique number (a very long one) for every computer program, whatever the programming language. The same encoding scheme can be used with each possible input set. Now, if we could build "the Oracle," its behavior could be described by Table 4.1.

Table 4.1 represents the internal decision logic of the Oracle by depicting each possible program as a row, every possible set of inputs as a column, and the corresponding outputs from the Oracle. The "☺" symbol indicates that the program will halt on the input and the symbol ☹ indicates that the program will not halt on the corresponding input. For example, the table shows that for Program 1 with Input Set 1, the Oracle would decide "☺," that is, Program 1 will halt on Input Set 1. Imagine that you have completed Table 4.1, that is, you have accounted for every program and input set in the Oracle's logic ("☺" or "☹").

You should see, however, that no matter how hard you try, a new program can always be created whose behavior is different from any other program already accounted for by the Oracle. This new program is created such that its output corresponds to the last row in Table 4.1. That is, if Program 1 outputs a "☺" for Input Set 1, then the new program outputs a "☹" for Input Set 1. If Program 2 outputs a "☺" for Input Set 2, then the new program outputs a "☹" for Input Set 2 and so on. If Program n outputs a "☹" for Input Set n, then the new program outputs a "☺" for that input set and so on down the diagonal of

TABLE 4.1

Cantor's Diagonal Argument to Show That the Halting Problem Is Undecidable

	Input Set$_1$	Input Set$_2$	...	Input Set$_n$	...
Program$_1$	☺	☺	...	☹	...
Program$_2$	☹	☺	...	☹	...
.			...		...
.			...		...
.			...		...
Program$_n$	☺	☺	...	☹	...
.			...		...
.			...		...
.			...		...
A new program	☹	☹	...	☺	...

the logic table to the very last program that the Oracle knows about. Since you can never complete the table the Oracle's logic is always incomplete, implying that the Halting Problem is undecidable.

Notice how in this tutorial I have combined narrative text with a figure and a table to illustrate a rather difficult concept. This approach is consistent with my preference for avoiding long runs of sentences.

Tutorials are difficult to write; you may want to describe and define everything because you fear that the reader may be a novice, or you may skip over important details that appear to you to be obvious. Have several readers review your tutorial writing, more than you would have review other kinds of technical writing; it will improve the presentation.

4.4 Opinion

Technical opinion papers are found in office communications or consulting when, for example, you are asked to endorse some new tool, technology, or methodology. Opinion or position papers are fun to write because you have license to use humor, anecdotes, metaphors, and other kinds of writing devices in addition to logic and precedent to support your position. I have written many opinion pieces and I enjoy writing them. I especially like having fun with the titles so that I can grab a prospective reader's attention.[3] Here are a few of the more amusing titles of opinion papers that I have published:

- "Another Ode to Paranoia"
- "The Joy of Spam"
- "The Burning Bag of Dung and Other Environmental Antipatterns"
- "Staying Clear of Boiling Frog Syndrome"

While you have great leeway in writing opinions, you need to be careful not to become careless. You can't make unsubstantiated claims; and even if you are being speculative, you must acknowledge that you are only giving your opinion. You cannot use hyperbole, and you must use humor with great caution—or not at all. I would not write something humorous to a client, even in an opinion piece.

Here is a deconstructed excerpt from a piece I wrote critiquing a practice used in Agile software development—the stand-up meeting. Stand-up meetings are also found, although not always by this name, in many other technical disciplines [Laplante 2003] (© 2003, ACM, Inc. Included here with permission).

Stand-up meetings are an important component of the "whole team," which is one of the fundamental practices of extreme programming (XP). According to the Extreme Programming Web site, the stand-up meeting is one part of the rules and practices of extreme programming: Communication among the entire team is the purpose of the stand-up meeting. They should take place every morning in order to communicate problems, solutions, and promote team focus. The idea is that everyone stands up in a circle in order to avoid long discussions. It is more efficient to have one short meeting that everyone is required to attend than many meetings with a few developers each [1].

Note that in the article all claims are substantiated, although in this excerpt, reference [1] ["Daily Stand Up Meeting"] is not shown. Further on in the article, I try to point out negative aspects of stand-up meetings by way of comparison with other unhappy stand-up activities:

I think there is something wrong with a meeting held standing up. Standing is inherently onerous. It is used for punishment in schools and in the military: "Stand in the corner," "Stand at attention," etc. Standing has an implicit element of authoritarianism and theory X control (which asserts that people will respond only to threats).

Then I offer a personal anecdote to make a point:

I endured regular stand-up meetings for three years. What made the meetings most painful was my boss. His main reason for the stand-up meeting was not to increase efficiency or embrace XP as much as it was to shorten human interaction beyond anything directly related to the work product. This is the same boss who never took me out to lunch because he believed it was a waste of time.

Finally, I close with a bit of logical reasoning:

For a methodology that emphasizes people over process, stand-up meetings seem contradictory: Don't structure meetings because informal communication is the best; hold highly structured meetings that discourage informal communication.

I have described several objections to stand-up meetings, and although some of my alternatives can also be dismissed, my hope is that those who embrace XP—or consider embracing it—will understand that the stand-up meeting is not necessarily what it seems. I suspect that many team members consider this aspect of the approach the most onerous. Perhaps alternatives should be embraced as part of the XP culture. After all, one of the tenets of agile methodologies is to "embrace change."

This editorial uses a combination of comparison, personal experience, logic, and scholarly reference to build a case. I like using multiple techniques in my opinion pieces, but you may use only one or two.

4.5 Research Papers

4.5.1 Survey of the Field

You may be asked to conduct background research on a technical area and summarize your findings in a report. Your report may also include observations about gaps in prior research and future research trends. The research report differs from the technical report only in that the former is intended for experts to help map the way ahead for future research, while the latter is intended as a primer.

There are several ways to organize your findings. One way is as a dry summary of papers and their contents. Another way is to organize the papers into a historical context, showing trends or epochs in the evolution of the technology. As with the technical report, you can also organize your research report as a set of research questions. Yet another way is to use an existing taxonomy or create your own—a structured decomposition of the field.

For example, here is an excerpt from one such taxonomy-based approach for a real-time image processing research [Laplante 2002]. My goal in this paper was to review the set of publications in the journal *Real-Time Imaging* for the previous seven years and to do a gap analysis, that is, to identify where I thought more research was needed. The report begins:

> To understand the present, we must study the past. To see just how far has the sub-field of real-time imaging advanced, a keyword search was done on various research databases on the keywords "real-time" conjoined with "imaging" for 25 years before the founding of the *Real-Time Imaging* journal (1995) and for the years since its founding. Table 4.2 summarizes the results of these searches. [Laplante 2002]

I gave the findings of my background search in summary form, as shown in Table 4.2.

Rather than give a dry summary of every paper found (there were more than 100), I listed the titles in a very long reference list, and clustered those

TABLE 4.2

Search Results for Various Research Databases on the Key Words "Real-Time" and "Imaging"

Database	Hits in Publications before 1995	Hits in Publications 1995 until January 2002	Total Publications	% Increase <1995 to =2002
INSPEC	322	220	542	−31.68
IEEE Explore	196	349	545	78.06
Academic IDEAL	31	832	863	2583.87
American Institute of Physics	228	48	276	−78.95
Kluwer Online	0	10	10	∞
ACM Digital Libraries	0	323	323	∞
JSTOR	219	98	317	−55.25
Elsevier Science Direct	152	185	337	21.71
Springer	0	131	131	∞
Information Science Abstracts	0	5	5	∞
Wiley Interscience	1	77	78	7600.00
Total	1149	2278	3427	98.26

Source: From Laplante, P. A., *Real-Time Imaging,* 8(5), 413–425, 2002. With permission.

papers by research areas identified by authors in previous works. The space of image processing was thus organized accordingly:

Image compression

Remote control and sensing

Image enhancement and filtering

Advanced computer architecture

Computer vision

Optical measurement and inspection

Scene types

Models

Image data types

Processes

I then proposed a modification to the list based on my own analysis:

> The proposed changes [are] as follows. First, the classification "image compression" has been combined with "image data types" to form the area "image compression and data representation" to recognize the fact that in real-time imaging, the main issue for data representation is to reduce storage requirements and transmission times.

Next, the classification "scene types" is combined with "models" and renamed to read "multimedia/virtual reality." Such a classification would include multimedia systems, virtual reality, scene modeling and real-time rendering related research.

Next, the term "intelligence" is added to the area of "computer vision" to more clearly capture the various soft-computing techniques that are involved in this area.

Then "computer vision" and "optical measurement and inspection" are combined because the only apparent distinction is that "computer vision" pertains to generalized schemes, while the other can be assigned to specific application domains, such as textile or agricultural inspection.

Finally, the classification "processes" is split into "algorithms" and "software engineering issues." This is due to the belief that it is important for real-time imaging researchers and practitioners to begin to focus on best practices of software. Perhaps because many imaging engineers are not trained in software engineering, or because of pressures to complete the project, basic software engineering practices are often not followed, or followed poorly. This situation seems to exist in both industrial and academic R&D labs.

The analysis led to the following proposed taxonomy for real-time imaging research:

Image compression and data representation

Remote control and sensing

Image enhancement and filtering

Advanced computer architecture

Computer vision and intelligence

Multimedia/virtual reality

Algorithms

Software engineering

Then I analyzed the set of published papers that had appeared in *Real-Time Imaging* to show how the distribution was uneven across the areas proposed in the new taxonomy. These observations led to the penultimate conclusion suggesting new research:

Relatively little research has appeared in image compression and data representation (8%) and in image enhancement and filtering (7%). Disappointingly, even less work has been published on remote control and sensing (<1%), and on software engineering for real-time imaging systems (1%).

Therefore, more work in these areas should be encouraged.

The approach I took in the *Real-Time Imaging* survey paper was a rather straightforward one, and an approach that you can use in your own work. The steps are simple:

1. Assess the state of current research by organizing existing papers into a known taxonomy.
2. Analyze the set of organized papers to identify any gaps.
3. Propose new research to be conducted based on the gaps.

You can propose any adjustments to the known taxonomy as warranted.

4.5.2 Based on Survey Data

Survey-based research is often difficult to conduct because of the challenges in getting a suitable number of respondents. The mechanics of conducting such research are outside our scope here. Let us instead consider the organization of the written report that summarizes the findings that may result from survey research. Technical survey-type research is very similar to marketing research; if you want to write this kind of paper in your place of work, you would do well to consult with your marketing department for some advice. Marketing professionals are experts at this sort of research writing.

As an example of this type of writing, consider a survey on requirements engineering practices that I conducted with a colleague (portions reprinted, with permission, from Colin J. Neill and Phillip A. Laplante, "Requirements Engineering: The State of the Practice," *Software*, 20(6), 40–46, 2003, © 2003, IEEE).

The introduction to the paper describes how we organized our survey. After describing the need for the survey, we explain the administration process:

> We created a Web-based survey (www.personal.psu.edu/cjn6/survey .html) consisting of 22 questions (summarized in Table 1). We drew our survey participants from a database of prospective, current, and past graduate students of the Penn State Great Valley School of Graduate Professional Studies. We sent them an email invitation (and subsequent reminder) to visit our Web site.

You can also describe a survey sample population graphically, as we did in another paper based on the same survey data (see Figure 4.2).

Returning to the survey, we described the survey questions in tabular form (shown in Table 4.3).

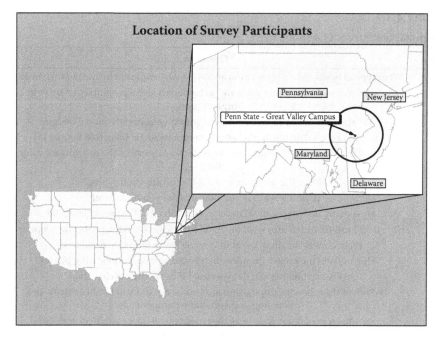

FIGURE 4.2

Penn State Great Valley School of Graduate Professional Studies location and service area. (From Laplante, P. A., Neill, C. J., and Jacobs, C., *Proc. 27th NASA/IEEE Software Engineering Workshop*, December 2002, pp. 121–128. With permission.)

The paper then describes an important component of any survey research, the response statistics:

> We collected survey data through March and April 2002, and although we received a few additional responses after this date, we didn't include them because the analysis had already commenced. So, of the 1,519 invited persons, we had 194 completed responses.

There are many ways to report survey results, but it is common to use bar and pie charts as in Figure 4.3. Here we reported on the survey results for the question "Which of the following development life cycles best describes the one you are using/did use?" The x-axis indicates the primary choices selected by survey respondents, and the y-axis indicates the percentage of respondents who selected that response to the question. The notation "$n = 191$" indicates that the number of respondents to the question was 191.

The main result of this survey—that more than 30% of respondents still used the Waterfall life cycle model—yielded several other papers and this

TABLE 4.3

Summary of Survey Questions

No.	Question
1	What type of business or organization are/were you employed by during this project?
2	Approximately how many software professionals are/were employed by your organization?
3	What is/was the approximate size of your organization's annual budget?
4	Which of the following application domains does/did this project apply to?
5	Which of the following development life cycles best describes the one you are using/did use?
6	Within the life cycle, do/did you do any prototyping?
7	If your answer is yes, how do/did you prototype?
8	How many full-time staff (IT) are/were involved in the project altogether?
9	How many full-time staff are/were involved in each phase of the project development?
10	What is/was the duration of the project (from inception to delivery)?
11	How is/was the project duration distributed among the following phases?
12	What techniques do/did you use for requirements elicitation?
13	Which of the following approaches are you using/did you use in analysis and modeling of the software requirements?
14	In what sort of notation is/was the requirements specification expressed?
15	Do/did you perform requirements inspections?
16	If your answer is yes, which technique do/did you use?
17	In your opinion, does your company do enough requirements engineering?
18	When you review/reviewed requirements, which of the following approaches do/did you employ?
19	Indicate with a number the size of requirements specification in terms of [the following].
20	The following statements are indicators for software quality and software productivity. Please rate these statements by clicking one box with the following scales.
21	Which of the following best describes your position while engaged in this project?
22	Over the last five years, how many software projects have you worked on?

Source: Neill, C. J. and Laplante, P. A., *Software,* 20(6), 40–46, 2003.

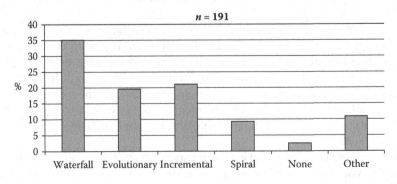

FIGURE 4.3

Reported life cycle model used. (From Neill, C. J. and Laplante, P. A., *Software,* 20(6), 40–46, 2003.)

result was reported in a prestigious international magazine [The Economist 2004]. Many other bar charts, such as Figure 4.3, appeared in the paper. The use of graphical elements, such as those just shown, is discussed further in Chapter 7.

4.5.3 Based on Experimentation

You may be asked to write a research report describing the results of an experiment or a set of experiments. This kind of technical writing is very straightforward and you can use the following format:

1. Describe the background and need for the experiment
2. Describe the methodology of the experiment
3. Report the results of the experiment
4. Interpret the results of the experiment
5. Offer recommendations and conclusions

The key to this kind of writing is to provide sufficient information to allow someone else to repeat the experiment. Readers may believe your results, but they will certainly lose faith if the results cannot be repeated. Even if no one chooses to attempt to repeat your experiment, providing ample background information increases confidence in the results.

To illustrate, I will describe an experiment conducted by one of my students and the subsequent technical report we published in a respected refereed magazine. In this case we wanted to explore the viability of Microsoft's C# language and .NET framework for use in real-time applications [Lutz and Laplante 2003].

After describing the various technologies involved in some detail, and some special concepts in real-time processing, the performance tests were described.

> We conducted two experiments comparing C#'s performance against C. All tests were built with version 55603-652-0000007-18846 of C++. NET, and we built and executed C#.NET on the.NET platform, version 1.0.3705. These versions correspond to.NET's first generally available, nonbeta version. All builds were optimized release builds. The tests ran on an 800-MHz Inspiron 8000, running Windows 2000 Professional, SP 1, with 523 Mbytes of physical memory.

The first test was the execution of 10 billion floating-point operations so we included a graph showing the performance results. We then described the second test, which dealt with memory management:

> We generated and released a linked list containing 5,000 nodes in both C and C# 24 times (48 total). Every other list saw an increase in node size. So for the first two sets of 5,000 nodes, each node contained a simple

numeric value and a string of length zero. For the second set of two lists, each node contained a simple numeric value again and a string of 2,500 bytes, and so on, increasing the string size by 2,500 bytes every other list iteration.

Note how we provide sufficient details of the experiment in order to allow interested parties to repeat the experiments. Not only are the computations explained, but also included are the characteristics of the computer on which the experiments were conducted.

Here is another example from a professional paper (thesis) written by one of my graduate students. The paper reports results of experiments with a new technique that I developed for dealing with various types of uncertainty in images [Rackovan 2005].[4] After some preliminaries, the paper describes the experiments:

> The image used for all of the experiments was the well-known "Lena" image (lena.jpg). All of the image filters were implemented with MATLAB and all executions were performed from the MATLAB command line. The images were converted to an array format, each containing a separate grayscale value ranging from 0 (black) to 256 (white). For ease of implementation, this image array was then transformed into a double floating point array ranging from 0 (black) to 1 (white), and then randomly contaminated with 3 different levels of uncertainty. These values of uncertainty were set to −3 (most uncertain), −2, and −1 (least uncertain). The contaminated image was then run through both versions of the mean filter, median filter, dilation filter, and erosion filter. Figure 4.4a shows the original image used in this investigation. Figure 4.4b shows the original image contaminated with uncertainty.

There is an interesting story behind the "Lena" images in Figure 4.4. This image has appeared in many books and hundreds of papers in imaging science journals since 1972. "Lena" is particularly useful to imaging scientists because it has both clearly defined and fuzzy edges and skin and hair tones

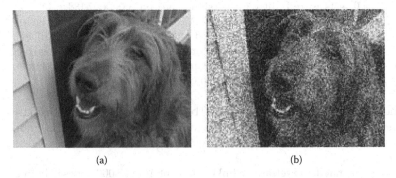

(a)   (b)

FIGURE 4.4
The Lena test image before (a) and after (b) adding noise.

that present interesting challenges for certain algorithms. Perhaps another reason for the persistent use of the photo is that Lena is attractive—in fact, the Lena picture is a cropped version of a frontal nude that appeared in the November 12, 1972, *Playboy* magazine. The pervasive use of this picture in image processing papers for so many years has created two ethical dilemmas—one relating to the "objectification" of women, the other to copyright infringement—most authors fail to obtain permission to use the image.

In 1992, Playboy Enterprises challenged the editors of one publication, *Optical Engineering*, alleging copyright infringement, but left the matter unresolved [Hutchinson 2001]. Despite these issues, Lena still appears in imaging papers, usually without proper attribution. Very recently I reviewed a paper for an image-processing journal that had an unattributed Lena image. Of course, Lena also appears in this book.

At this point you must be confused because Figure 4.4a and 4.4b do not depict a woman. Figure 4.4a and 4.4b are actually photographs of my dog, Teddy (sadly, he passed away a few years ago). I requested permission from Playboy Enterprises to use the image of Lena, but after several requests, I received no response. So, I had to use a substitute—and Teddy gave me permission to use his likeness. If you want to see the real Lena picture, just Google it.

Getting back to the thesis, another feature is the representation of an algorithm in a programming language, in this case, pseudocode. Pseudocode is a generic name for any code syntax that resembles a programming language but is not intended to be compiled and executed. The idea is that if the reader understands C, Fortran, Java, or some other modern programming language, then they would get the general idea of the algorithm by inspecting the pseudocode. Here is the introductory text and the accompanying pseudocode:

> The algorithm used for contamination of the original image is shown in Algorithm 1. A random pixel array was generated using the MATLAB rand function. This function generated a 512 × 512 array of random numbers ranging from 0 to 1. This array was traversed and the original lena.jpg was given values based on the logic displayed in Figure 4.5.

```
if random_map(i,j) > 0.98
        Iunc(i,j)=-1;
elseif (random_map(i,j) > 0.96) & (random_map(i,j) <=0.98)
        Iunc(i,j)=-2;
elseif (random_map(i,j) > 0.94) & (random_map(i,j) <=0.96)
        Iunc(i,j)=-3;
else
        Iunc(i,j)=Iorig(i,j);
end
```

FIGURE 4.5
Algorithm 1 contamination of original lena.jpg.

Of the 262,144 pixels in the contaminated image, 5,220 (1.991%) had an uncertainty value of −1; 5,242 (2.000%) had a value of −2; and 5,231 (1.995%) had a value of −3. This contaminated image served as the standard input to all the enhancement filters used in this investigation.

The key point here is that when expressing algorithms, technical writers should use pseudocode because readers shouldn't have to learn a new programming language to understand your experiment or conclusions.

In his thesis, as is often the case with technical writings, my student generated many numeric results. These results are most easily summarized in a table, preceded by an introduction. Here is how he handled this situation:

> Based on the results from all of the filtering methods, it can be determined that the EFPS filtering operations perform better than Regular filtering operations when applied to an image randomly populated with uncertainty. More specifically, all of the EFPS filtering methods reduced the number of corrupt pixels in the resulting image while retaining the intended functionality of the specific filtering operation. The EFPS Mean filter proved to be the most effective, and successfully removed all uncertainty from the originally corrupted "Lena" image (Figure 4.4b). Conversely, all of the Regular filtering methods resulted in images containing more corrupt pixels as compared to the original image.
>
> Table 4.4 summarizes the performance of all of the Regular and EFPS filtering techniques. The numbers and percentages displayed correspond to the amount of uncertainty present in the resulting image before and after it was filtered:

TABLE 4.4

Summary of Filter Performance

Filtering Technique	Uncertain Pixels: Pre-Filtering			Uncertain Pixels: Post-Filtering		
	−3	−2	−1	−3	−2	−1
Regular Median (8-Neighborhood)	5231 (1.991%)	5242 (2.00%)	5220 (1.995%)	42892 (11.621%)	36654 (13.982%)	30864 (16.362%)
EFPS Median (8-Neighborhood)	5231 (1.991%)	5242 (2.00%)	5220 (1.995%)	0	3(0.001%)	20 (0.008%)
Regular Mean (8-Neighborhood)	5231 (1.991%)	5242 (2.00%)	5220 (1.995%)	31009 (11.829%)	37266 (14.216%)	42793 (16.324%)
EFPS Mean (8-Neighborhood)	5231 (1.991%)	5242 (2.00%)	5220 (1.995%)	0	0	0
Regular Morphological Filter	5231 (1.991%)	5242 (2.00%)	5220 (1.995%)	14163 (5.403%)	14769 (5.634%)	14984 (5.716%)
EFPS Morphological Filter	5231 (1.991%)	5242 (2.00%)	5220 (1.995%)	1(~0.0%)	14(0.005%)	34 (0.017%)
Regular Morphological Dilation	5231 (1.991%)	5242 (2.00%)	5220 (1.995%)	13738 (5.241%)	14848 (5.664%)	15369 (5.863%)
EFPS Morphological Dilation	5231 (1.991%)	5242 (2.00%)	5220 (1.995%)	0	18(0.007%)	44(0.017%)

Finally, he interprets the results:

> Many recommendations arise from the strong assumptions made in this investigation. All uncertainty generated randomly was given the values of –1, –2, and –3 (or 2, 3, and 4 for the morphological erosion filters). The ability to classify these varying levels of corruption remains undefined and may lead to extremely subjective classifications. Based on this prospect, the filtering algorithms presented in this investigation are useless without extensive pre-filtering characterization of the possible types of uncertainty residing within the corresponding image data. Numerous techniques have been adopted that identify various types of image contamination. An extension of this investigation would be to consolidate past works in assorted areas of uncertainty classification and implement these results as the pre-filtering procedure to all of the EFPS techniques, specifically the population of the fuzzy set used to define the varying levels of image contamination.

These excerpts from my student's thesis are great because they showcase several elements: graphics, tables, code snippets, and experimental data. The Lena story is a good teaching moment, highlighting the challenges in procuring copyright permissions and focusing on working around setbacks.

4.6 Reviews of Books, Papers, and Reports

4.6.1 Reviews

Reviews of technical books, research papers, and reports take a form similar to reports of nontechnical works. You may be asked to write such reviews for your employer, for a client, or for a class. Therefore, it is worth understanding how to write good reviews.

I have served as an area editor for ACM's *Computing Reviews*, which is the leading forum for reviews of books, conference papers, journal papers, and other scholarly writing on the computational sciences. In this capacity, and as a professor, I have reviewed hundreds of other peoples' reviews that range from exquisite to sophomoric and even worse. I have also written and published a few reviews myself. Based on these experiences, here is some advice on writing good reviews for various types of publications. Should you have to write a review for a technical paper or book, I have included a template in Appendix B.

4.6.2 Journal and Conference Paper Reviews

Journal and conference paper reviews are straightforward: You want to capture the essence of the work and its importance, identify any shortcomings, and then make a recommendation as to the type of reader who would be most interested in the paper. Here is a detailed outline of a review. The first and last elements are essential; the other elements can be used or omitted as appropriate.

1. Grab the reader's attention in the first sentence.
2. Set the context for the work being reviewed:
   a. Why is this work important?
   b. How does this work fit with respect to related work?
   c. Is there something important about the authors to note?
3. List the salient findings or results of the work.
4. Discuss any shortcomings of the work, and suggest possible improvements.
5. Recommend areas for future research related to the work.
6. Recommend the type of reader who would most benefit from the work (e.g., researcher, practitioner, novice, or expert).
7. Recommend further readings.
8. Give your final assessment of the work and any salient conclusions.

Here are some excerpts from a review that I wrote for a paper about software improvement [Laplante 2006] (© 2006, ACM, Inc. Included here with permission). First, I try to get the reader's attention:

> An old saying goes, "When you don't know where you are going, all roads will take you there." The authors would have you reinterpret that saying as follows: "If you know where you are going, regardless of how much time and money you waste, how many accidents you have, and how much you hate the ride, all that matters is that you become a better driver"; In short, this piece seems intended to salvage the reputations of those who led a disastrous software process improvement (SPI) initiative.

I then proceed to describe the experiment that was conducted involving an anonymous company and a particular process improvement framework called the Capability Maturity Model (CMM). I recount particular details of the company in question, and note that certain details were omitted. For example:

> Although employee surveys were used extensively during the initiative to solicit feedback, we don't know much about this feedback (though we are told there were "no major problems with the methodology or with reluctance by the developers in using it"). But it is hard to believe that the response was overwhelmingly positive. In any case, it seems that a serious flaw in this approach was that no appropriate buy-in or input was obtained from developers prior to the mandate—only after its pursuit was a fait accompli.

I wrap up by recounting the nine lessons that the authors believe were learned from the experiment: I note that these lessons are already well known. I then give a rather scathing critique of the research:

> [The conclusions sound] like the CIO's damage control statement to the board and shareholders. And, although the authors allude to the benefits that were achieved by the attainment of level two (in the areas of customer satisfaction, number of days of deviation from scheduled delivery date, and the percentage of deviation from the proposed budget), we are given no data. Indeed, from the way the narrative is written, it is unclear that any of these measures improved at all.

Finally, I note that too much information is missing to repeat this research and that the results seemed to have been spun to support the foregone conclusions.

> The authors conclude that "software process improvement is a journey, not a destination." Although this sentiment is correct, it shouldn't be used as an excuse for a failed initiative, particularly when the cost of learning such obvious lessons is so high. By choosing to act as apologists for the CIO of AAC, the authors miss the real lesson of this case.

Clearly, my opinion of this paper was not very high, and the review is a negative one. But the structure of a positive review would be the same.

4.6.3 Book Reviews

Good book reviews are very hard to write—much harder than reviews of papers. The best reviews focus on themes that tie the book together. This is not always easy; for example, academic texts are less likely to have deep underlying themes. But the job of the reviewer is to uncover a hidden structure that goes beyond what one can get by reading the table of contents.

The review should weave the narrative around the book's central theme to reveal its essential concepts. The review should not consist of a litany of chapter summaries. For example, consider this snippet of a deliberately poorly written review, for the book: Frederick Brooks' *Design of Design: Essays from a Computer Scientist*, Addison-Wesley Professional, 2010.

> Brooks' *Design of Design* is a master work from one of the leading computer scientists in history. This 448 page book is an interesting read from cover to cover. The book's 28 chapters are organized into six sections....
>
> The book starts with Section 1: Models of Designing, consisting of five Chapters. Brooks gives us his list of issues and paradigms for design....
>
> In Section 2: Collaboration and Telecollaboration, Brooks tackles the issue of the modern distributed workplace....
>
> In Section 6: Trips Through Design Spaces, Brooks provides us with seven case studies of design from his own experience, and concludes

with "Recommended Reading," which provides a nice list of references for the interested reader.

This is one of the best books that I have ever read. Everyone should have "Design of Design" on their bookshelf.

Why is this review so bad? It starts frivolously: "master work" is a meaningless platitude. The second sentence is worse. The length of the book is unimportant (you could look that up on Amazon.com) and the sentence ends with a cliché. Next, we are led through a dry rendition of the book's contents, each paragraph fitting the same template ("In section x"). The review concludes with a predictable and boring conclusion. In my capacity, as one of three software engineering editors for *Computing Reviews*, I have unfortunately read far too many book reviews by experienced professionals, senior scientists, and professors that were just as poorly written.

To contrast, here are some excerpts from the real review that I published for this book in *Computing Reviews* [Laplante 2010] (© 2010, ACM, Inc. Included here by permission):

> Best known for his seminal work on software engineering and project management, Frederick Brooks expands here to ubiquitous design. Having led the development of the IBM 360 computer and its operating system, I suspect that many do not know about his pioneering work in virtual reality. Brooks draws upon both domains, and more pedestrian endeavors such as the renovation of his own home, to seek universal meaning in physical design and the design of software.

In this opening paragraph, I am trying to draw the reader's interest. Then I use the fact that the book has a theme around "spirals" to lead the discussion.

> Brooks likes spirals. There is the Boehm spiral model, which is mentioned in several of the essays, a picture of a spiraling bridge, and even a spiral staircase in the author's beach house.

I then highlight a few of the chapters' contents, but not all of them. I conclude the review with the following:

> This book is many things, but mostly it is a look into the mind of one of the greatest designers of the automated age. If I were to make a movie based on this book, it would mimic *Being John Malkovich*, where the protagonist gets into the head of a famous American actor. Being Frederick Brooks would be a frenetic movie—thrilling at times, distracting at times, and puzzling at times—but there would be so much to be learned. I suggest that every computer scientist and software engineer buy this book.

Note how, despite my warnings about being funny, you can sometimes work humor into a review.

4.6.4 Blind Reviews

If you are an expert in your field, or simply a typical reader of some magazine, journal, or other periodical, you may be asked to review, for potential publication, the work of another author. In this case you are acting as a very special kind of reviewer or "referee." Your identity is protected from the writer (this is called a "blind") so that you may be honest without fear of retribution, argument, or harassment from the authors. In some periodicals, the identities of the authors are hidden from the reviewer during the review process so that any personal biases will be mitigated. In these cases, the review is known as "double-blind" because both the reviewer and the authors are unknown to each other. Only the editor knows the identities.

The editor of the periodical will take into consideration the reviews of the referees (usually at least two and sometimes as many as six) and render a consensus judgment. The usual judgments can be "accept without change," "accept with minor changes," "revise the paper and resubmit it for more review," and "reject." Papers can also be returned without review if they are deemed to be "out of scope," that is, inconsistent with the stated purposes of the periodical.

As a referee, your goal is to fairly, but critically, review the paper for the purposes of improvement. In the review, you are giving advice to both the authors and the editor. You need to put your personal feelings aside and be as objective as you possibly can be. In the worst cases, however, where the work is weak or very badly done, you may have to recommend rejection of the paper.

Here is an example of a blind review that I wrote for a short and informal paper that discusses a technique used in software engineering (and other kinds of engineering) called the "design interview." The paper had been submitted for a special issue of a magazine with the theme of "tools."[5]

> This paper is a little off-base from the tools theme issue. The paper introduces a methodology, or really, a micro-methodology. I like what the author is proposing, but I don't think we can call his proposed methodology a tool.
>
> This paper introduces a micro-methodology, the design interview. It's not a methodology in the sense that it is a fine-grain technique that can be used at every stage of the software life cycle. In my mind, the design interview does more than elicit information about progress. It's a team-building mechanism. It has to have a positive effect on the organization, team and ultimately the project, if the manager is caring and positive.
>
> When talking about embraceable techniques, I wonder what the effects of the lightweight methodologies (like XP or Scrum) have on embraceability? That is, it is reported that programmers tend to like to work this way as opposed to the traditional command and control, waterfall paradigm. Maybe a discussion on this might be fruitful.

I clearly liked this paper and could only offer a short recommendation for improvement. I left the decision to the editor as to whether this paper was within the scope of the special issue theme.

Some editors like to have two sets of comments from reviewers: one set for the authors and then a confidential recommendation to the editor. This approach is particularly convenient when the review has some negative elements. Here is an example of a review for a paper on project management that I wrote for another magazine. First, I provide comments to the author:

> I like this paper, and it addresses an important issue, but I found it somewhat lacking in specifics. In general, I think my main criticism is that it is a bit too abstract—while I understand the need to protect proprietary data, I think some more quantitative information is necessary. For example,
>
>> What are some industry standards for "repeat business" and "strong referral" (mentioned on page 2). Then, perhaps you can talk about a percentage increase in these figures without giving away confidential information.
>> The same goes for the Pursuits and Assessments and Implementation factors on page 3 (e.g., Win rate, project margins, client satisfaction, and timely payment). Talking about these in the abstract is less helpful than getting real numbers—even if they are simply industry standard numbers.
>> On page 5, are "red projects" good or bad projects? In that same paragraph you mention a timesheet, a snippet example of one might be helpful.
>> On page 6, the "Top 4–7 risks are examined in detail." It might be nice to have a table with a sampling of some of these.
>> On page 6/7, "best practices are harvested...)," a short list of these might be nice.
>> On page 8, maybe a few excerpts of sample audit responses would be helpful.
>> On page 10, under "Did it Work?" I realize that the repeat business% is confidential, but perhaps the improvement% of repeat business would be nice. If this approach, e.g., led to a 10% increase in repeat business, companies would be scrambling to implement this!

Notice how I try to provide constructive criticism without attacking the author. Next, I give some advice:

> I don't know if the author got trapped by the word count limitations, but I think that more empirical data would give the paper much more credibility.

In the comments to the authors, I include very fine details as well as give my overall impression of the work. As a reviewer, you are expected to give both macro-level guidance and identify typographical errors, errors in equations, and errors in concept, as well as identify missing information.

4.6.5 Vignette: Scientific Proposal

To conclude the chapter, here's an excerpt from a proposal for a small grant project that I applied for many years ago. Proposals are discussed further in Chapter 6, but this vignette provides another example of scientific writing:

> A system Test Plan will be written that will incorporate hardware, software and integration testing. Hardware testing will cover the entire development life cycle including validation of system requirements using Requirements Reviews, model checking and consistency checking where appropriate; design validation using Design Reviews and simulations; hardware testing using best practice and final system integration testing via long term operational performance under load via fault injection.
>
> Typically, Field Programmable Gate Array (FPGA) testing consists of successively configuring the FPGE using a set of configuration inputs and then applying a test sequence for all configurations using the appropriate operational limit. After manufacturing, the complete FPGA will be tested independently of the application and again with the application installed using pre-defined test scenarios, which will include injected faults.
>
> The detectability of hardware faults in FPGAs depends on the circuit configuration. A given fault may be redundant (undetectable) in some configurations, while non-redundant in other configurations. Therefore, design for test requires that an appropriate configuration be identified, which will minimize the possibility of undetectable faults.
>
> Finally, built-in software testing specifications for the hardware will be generated so that ongoing software testing of the hardware subsystem can be performed.

The project was not funded, but the excerpt provides a nice example of scientific writing for discussion or editing (it can be improved).

4.7 Exercises

4.1 Write a short (no more than two pages) tutorial on your favorite hobby.

4.2 Write a product review for an electronic device of your choosing. The review can be similar in form to those found on Amazon.

4.3 Select a scholarly paper of interest to you (use Google Scholar) and write a review. Keep your review around 200 words in length.

4.4 Write a review for a technical book of your choosing. You can use the template for a review in Appendix B. Book reviews should be between 500 and 800 words in length.

4.5 Prepare a sample permission request letter to reprint a portion (e.g., a 200- to 300-word excerpt) from the book you reviewed in Exercise 4.3 in another work that you are preparing (e.g., a research paper).

4.6 When is a chronologically based review of previous research called for, and when would this be a poor choice for organizing such a review?

4.7 Consider three questions to guide the evaluation of a scholarly paper:

   a. What is the conclusion?

   b. Is it so? (In other words, is the conclusion well supported?)

   c. So what? (Is there any significance to the findings?)

   Do you think that these three simple questions comprise an adequate framework for evaluations of scholarly papers? Why or why not? What other criteria would you add?

4.8 In the opening paragraph of the review that I wrote for Frederick Brooks' book, *The Design of Design*, there is a problem with the second sentence "Having led the development of the IBM 360 computer and its operating system...," particularly if you ignore the previous sentence. What is the problem with the second sentence? (Note: This example illustrates that even a fully edited and published writing can contain errors).

4.9 For the Scientific proposal vignette in Section 4.6.5, identify the jargon words and rewrite the proposal without them.

4.10 For the Scientific proposal vignette in Section 4.6.5, rewrite the proposal to improve the writing. You can retain the jargon words or remove them as you did in Exercise 4.9.

Endnotes

1. Here is a "permission teaching moment." The ACM provides the author the "right to reuse any portion of the work, without fee, in future works of the author's own, including books, lectures and presentations in all media, provided that the ACM citation and notice of the Copyright are included" (http://cacm.acm.org/help/copyrights-permissions/).

2. © ACM, 2008. This is the author's version of the work. It is posted here by permission of ACM for your personal use. Not for redistribution. The definitive version was published in Laplante, P., "Open Source: The dark horse of software," *Computing Reviews*, July 15, 2008.

3. In a moment of mischievousness, I embedded a four-letter swear word noticeably in a paper that was to be published. Feeling guilty, I notified the editors of the prank but they let it stay in the published work [Laplante 2006].

4. Here is a different permissions situation. In most cases, when a student submits a dissertation or thesis to a professor, he signs a paper transferring copyrights to the university. But the university authorizes the supervising professor to reuse this work as long as the proper citation is made.

5. Here is another permissions interlude. Because I wrote this review and never signed away the copyright, I own the rights to it. And I have disguised the review so as not to give away the identity of the publication or authors.

References

"Daily Stand Up Meeting," http://www.extremeprogramming.org/rules/standup meeting.html.

The Economist, Measuring complexity, November 25, 2004.

Hutchinson, J., Editorial, Culture, communication, and the information age, *IEEE Professional Communication Society Newsletter*, 5(3), 1, 5–7. 2001.

Laplante, P. A., A retrospective on real-time imaging, a new taxonomy and a roadmap for the future, *Real-Time Imaging*, 8(5), 413–425, 2002.

Laplante, P. A., Stand up and deliver: Why I hate stand-up meetings, *Queue*, 1(7), 7–9, 2003.

Laplante, P. A., The joy of spam, *Queue*, 4(9), 54–56, 2006.

Laplante, P., Open source: The dark horse of software, *Computing Reviews*, July 15, 2008.

Laplante, P. A., Review of *The Design of Design: Essays from a Computer Scientist*, 2010, Addison Wesley by Frederick Brooks, in *Computing Reviews*, April 2010.

Lutz, M. and Laplante, P. A., An analysis of the real-time performance of C#, *Software*, January/February, 74–80, 2003. Reprinted in Distributed Systems Online, dson line.computer.org/0302/f/sp1lap.htm.

Neill, C. J. and Laplante, P. A., Requirements engineering: The state of the practice, *Software*, 20(6), 40–46, 2003.

Rackovan, A., Handing of Uncertainty in Image Processing Applications, MS thesis, Penn State Great Valley, Summer 2005.

5

Business Communications

5.1 Introduction

In addition to writing documents that are particular to your technical discipline, you will have to prepare memos, nontechnical reports, agendas, meeting minutes, business plans, résumés, cover letters, and so forth. Most of the topics in this chapter are generic. For example, a good résumé follows the same principles regardless of your profession. Your preparation of standard business communications will be greatly enhanced, however, by understanding several fine points that are particular to the technical professional.

5.2 Résumés

The résumé is perhaps the most personal of business communications. Your résumé and cover letter may be the two most important technical documents that you will ever prepare. An excellent résumé and cover letter can unlock a better job and increased earning potential.

I have reviewed hundreds of résumés for technology and business professionals as a project manager, technology executive, department chair, dean, and college president. I can assure you that a great résumé does not guarantee that you will get hired—you have to sell yourself during the interview and afterward. While a great résumé may lead to an interview, a bad résumé will surely preclude an interview.

Common elements of a résumé include (items in boldface type are essential):

1. **Name**
2. **Contact Information**
3. Summary
4. Statement of Objective

5. Experience

6. Education/Training

7. Licenses and Certifications

8. Consulting

9. Hardware and Software

10. Foreign Languages

11. Security Clearance

12. Military and Other Service

13. Awards and Honors

14. Publications

15. Affiliations

16. Interests

17. References

There are other possible areas to list in a résumé, particularly in highly specialized professions, but the aforementioned elements are common. Of course, if you have nothing to list under an area, omit the header—don't list the header followed by a blank line or "none."

Résumés should be short. I prefer a résumé that fits on one side of a sheet of paper, although for a long job history, that might be difficult to achieve. Hiring managers tend to get turned off by résumés that span more than two pages. You can always reduce the size of the résumé by deferring certain elements to separate attachments, for example, the lists of references, publications, hardware, and software. Let's explore some hints for preparing each element of a résumé.

5.2.1 Name

The main issue in listing your name is whether to be formal or informal. There is a difference between "Fred Blog" and "Mr. Frederick J. Blog, III." Informality connotes approachability and modesty, but it also can imply carelessness and laxity. Formality has the connotation of seriousness and power, although it also suggests vanity and detachment.

Listing your name informally is more appropriate for certain jobs, for example, technical support or installer. The formal version is more appropriate for a senior management position. When in doubt, or if the decision could go either way, I would opt for the formal name without a salutation: for example, "Frederick J. Blog, III."

I list my credentials next to my name (see my résumé in Section 5.2.21). You should list your credentials too if they are important to the job under consideration.

5.2.2 Contact Information

For contact information, you should include a physical address, a phone number, and an e-mail address. A surface address is necessary to establish your place of residence, but you do not need to give your home phone or work phone if these pose problems. For simplicity, you should not list multiple surface addresses even if you reside in several places. You may list multiple phone numbers (e.g., home, work, cell) and multiple e-mail addresses. If you list multiple phone numbers or e-mail addresses, be sure to clearly note which one is preferred for employment-related communications.

Do not give Facebook, Twitter, LinkedIn, or other cyberspace locations, unless for some reason, this is conventional in your area of employment (e.g., if you are a software game developer). Remember that where you reside and how you communicate offer clues about your work and life habits, and you want these to reflect favorably upon you in a professional context.

5.2.3 Summary

Some people like to give a snapshot of their skill set in a summary. Summaries can assist hiring managers to quickly determine if you are a possible match for a job. But the summary can also give a hiring manager an excuse to quickly dismiss you without reading your résumé in detail.

In cases where it would be too generic, I suggest omitting the summary. For example, the following is flabby:

> Almost six years of (extensive) programming experience in various languages with five years of extensive Web development (programming and design). Expert in MySQL database design.

The experience listed is quite good, but many other résumés with similar characteristics are likely to be submitted for consideration. In this case I think it would be better to forego the summary and let the details of these experiences be described in the "Experience" section.

In cases where your summary is profound, you should include it. For example:

> More than 30 years of experience as an engineering professional from junior engineer to vice president of technology for a Fortune 500 company. Earned PhD in electrical engineering, with over 300 publications and 27 patents in embedded telecommunications technologies.

There are probably less than a hundred people in the world who could claim similar experience, so this summary pops out. In most other cases, however, I suggest omitting a "Summary" section and include that information in the cover letter.

5.2.4 Statement of Objective

I don't advise putting a "Statement of Objective" in your résumé; but if you choose to include one, it should be focused. Any vague statement or one filled with platitudes (e.g., "I seek a position in which I can be a positive change agent") will likely relegate your résumé to the bottom of the stack.

The "Statement of Objective" gives the résumé a narrow focus. Instead, make your résumé as general purpose and comprehensive as possible, and save a statement of your career objectives for your cover letter. This way, you can tune the objective statement to fit the specific job for which you are applying without having to tweak the résumé.

5.2.5 Experience

Your experience is probably the most important component of the résumé so you need to be thorough, accurate, and honest in your reporting. Use action verbs to describe your real contributions. For example, the phrase "Part of a team that developed a $1 billion weapons system" says nothing about your role. Instead, you should be specific: "Wrote software acceptance test scripts and conducted acceptance testing for a $1 billion classified weapons system."

List all experiences in reverse chronological order, with your most recent experience listed first. In case of space limitations, include more detail about your most recent experiences, and less detail about older experiences.

5.2.6 Education and Training

Listing your dates of college enrollment and degrees earned is important. Do not list grade point average (GPA), even if you just graduated. Different people have very different reactions to a GPA. For example, while some may be impressed by your high GPA, others may assume that you are simply bragging. If employers care about GPA, they will require you to send official transcripts from any educational institutions that you attended.

Don't forget to list other education that did not lead to a degree. For example, professional training or certifications courses should be listed, particularly if they are relevant to the job. Training not directly related to the job is okay to list, but you need to weigh the value of that training. For example, if you have earned a purple belt (an indication of significant proficiency) in Brazilian Jiu Jitsu, you would not include that achievement in an engineering résumé. If you have some kind of specialized training that relates to the job indirectly, however, you do want to list that. For example, if you are working in a dangerous environment and you have life-saving training, then you ought to mention it.

5.2.7 Licenses and Certifications

Professional licensure through a State Board may be a requirement or preferred for many technical positions. Certifications, which are granted by a wide variety of entities including companies, user groups, or professional organizations, may also be required or preferred for a job.

Be sure to list any licenses and certifications that you have, even if these are not explicitly requested in the job advertisement. Licenses and certifications demonstrate individual commitment and achievement, and you never know how that extra license or certification could resonate with the prospective employer. For example, if you and the interviewer hold scuba diving certifications, this coincidence would reflect favorably upon you and provide a topic for informal conversation.

5.2.8 Consulting

Between periods of "permanent" employment, many technical professionals earn income and keep their skills current through consulting work. If you have a successful consulting practice, you can list your top clients and the engagement type in a separate section.

In cases where you may have been unemployed for long periods of time (say, more than three months), you can list "Independent Consultant" between your regular employment experiences and include a brief description of your engagements. This treatment for periods of unemployment is very common and shouldn't raise an issue with prospective employers. You may be asked for more details on those engagements during an interview, however. If those so-called "consulting engagements" have nothing to do with your profession, then they may diminish your credibility. On the other hand, if you can show a professional development angle for an unrelated engagement, then that may be a plus. For example, helping build homes for the homeless may not be a direct professional experience for an electrical engineer; but if you were the lead volunteer on electrical wiring installation in those homes, you might want to mention the volunteer activity.

5.2.9 Hardware and Software

You should list any specialized hardware and software that you have used during your career. You should list your proficiency for each of these—no one is an expert in everything. I suggest you use the following scale: proficient, moderately proficient, or familiar. Or you can list specific skills you have acquired with the technology (e.g., "capable of running full start-up and shut-down procedures"). List specialized training pertaining to hardware and software under the section on "Education and Training."

For example, it is customary in information technology (IT) and computing positions to list the programming languages with which you are familiar. Here is an example:

Computer Languages:

Proficient in PHP, HTML, JavaScript, jQuery (Ajax framework), VB.Net

Moderately proficient in CSS, C#, Ajax (using jQuery)

Familiar with JSON, .Net framework

Databases:

MySQL: Create, read, update, delete operations from PHP, multi-table joins, operators, built-in functions; table/database structure managed with phpMyAdmin

If you are not an IT professional, this list is probably meaningless to you, but the list would be helpful to an IT hiring manager in evaluating a candidate's skill set.

Use good judgment when including outdated technologies in your hardware/software list. Many positions involve old legacy systems, and proficiency in the use of these older technologies may be requisite. But I have seen résumés of electrical engineers listing proficiency with slide rules, ancient calculators, and 1980s home computers. Seeing a list like this suggests that the engineer is averse to change, or has not kept his education up to date.

5.2.10 Foreign Languages

Proficiency in a foreign language is a sign of sophistication. When you claim facility in a foreign language, be sure to describe your abilities in terms of reading, writing, and conversation. You can use the following scale: limited, working proficiency, full professional proficiency, native (or bilingual) [SIL 1998]. To officially be rated in these areas, you need to take a test administered by a reputable firm. But I think it is acceptable to self-assess if you are honest. Your interviewer may also be proficient in the language you claim and may test your abilities.

I have some working knowledge of French, but I choose to omit that fact on my résumé because I was once challenged in an interview and I don't think I fared well. If I did include the information, here is an honest self-assessment:

French: Reading (working proficiency), Writing (limited), Conversation (limited)

I think it is always good to list your skill with any foreign language, even if that language is extinct, like Latin, or one that is not widely spoken, such as Esperanza. But I wouldn't list novelty languages, for example, Klingon.

5.2.11 Security Clearance

You should list all active security clearances held. I would not list inactive clearances. Although there is some advantage in having held a past clearance because an agency only needs to investigate from the date of your last clearance to the present, you could mention your previous clearance status during the interview, if applicable to the position being discussed.

5.2.12 Military and Other Service

You should always list any military or civil service where you fulfilled a long-term commitment (e.g., the United States Peace Corps). The only possible exception is if you have some kind of less-than-honorable discharge. You will be required to disclose such a situation when you fill out a job application. But there is no reason that you must include that information on your résumé. A "dishonorable discharge," even if the circumstances are compelling, will hurt your chances of being invited to an interview.

5.2.13 Awards and Honors

You should list any awards or honors that are related to your career and also those that are not related to your career but are substantial. You should be able to provide evidence of all awards claimed (e.g., with a letter or certificate).

I can't give you an exhaustive list of appropriate awards, but here are few examples. Awards that are directly related to your career include accolades from professional organizations, significant recognition from your employer or from a customer, and recognition from any colleges and universities that you attended. Significant awards not related to your career, but which are worth listing, include earning the highest rank available as a Boy or Girl Scout, and receiving honors or awards for any kind of public service or humanitarian accomplishments. I wouldn't list sports awards of any kind unless they are truly significant; for example, you were a first alternate for the Olympic Fencing Team. You can discuss awards related to athletics or hobbies at an appropriate point during an interview.

Avoid listing vanity awards. There are a number of outfits that publish *Who's Who* books for the express purpose of getting the honorees to buy copies of the book. Don't fall for these traps; and if you did already, don't list the "honor" in your résumé. Such awards as "Selected for *Who's Who in Young American Professionals*" will be seen for the vanity it represents and will likely hurt your prospects. Obviously, omit such nonsense as "voted most likely to succeed in senior year of high school." What you list (and do not list) under your awards portrays a certain image of you.

5.2.14 Publications

It is always noteworthy if you have conducted organized research and published a paper about that work. Even one publication in a conference related to your profession may set you apart from many others. Therefore, list any publications that are related to your profession, including newspaper articles, magazine articles, journal papers, conference papers, and any other print or electronic venue where your work or opinion has been published. The format of the citation for a publication is the same as that for referencing discussed in Chapter 2.

In cases where it is common to have many publications, for example, if you are a research scientist or professor, or if you have a large number of publications because you like to write, list these in a separate document. I have seen résumés as long as twenty-five pages due to long publications lists, and to me, this reflects poorly on the job applicant's ability to concisely organize his presentation.

5.2.15 Affiliations

You should list your membership in certain organizations, learned societies, and honorific groups. I advise you to list only affiliations with professional organizations or groups that are relevant to your career. For example, if you are a mechanical engineer and you belong to the American Society of Mechanical Engineers, put that on your résumé.

Don't list your affiliation with any organization that has nothing to do with your career. Putting these irrelevant affiliations on your résumé might distinguish you in either a positive or negative way. For example, if you are the president of the local "Save the Whales" chapter, that might resonate with a prospective employer who shares your values, but will hurt your chances if the hiring manger does not share your priorities. You can discuss your personal interests and hobbies during an interview if it seems appropriate.

5.2.16 Interests

I don't think you should include a list of your hobbies, interests, groups, and clubs, etc. My rationale for omitting this information is the same as for "Affiliations"—you don't know if your interests will be viewed positively or negatively. The only exception is when your hobby is very directly related to your profession. For example, a microwave engineer might list hobby participation in amateur radio contests, especially microwave radio distance transmission competitions.

5.2.17 References

I don't advise listing professional references on your résumé. Instead, enclose a separate sheet of references with your application if they are required at that time. Customize your list of references for each job application.

Choosing and managing your list of references is very important. Here are a few tips on collecting your references:

> Always ask permission before listing someone as a reference.
>
> Always have more references lined up than you think you will need. In this way, you can select the best subset of your references for each job prospect.
>
> List references who can discuss your abilities and experiences.
>
> Try to list at least one reference for each job that you have held.
>
> References who still work at the company where you claim the experience are more effective than those who no longer work at that company.
>
> If you have held supervisory positions, list at least one reference of someone who works or has worked for you.
>
> For each work experience you list, you should have a supervisor as your reference, not a peer or subordinate. The exception is if there would be problem with listing the supervisor as the reference.
>
> Don't list relatives unless they were a direct or indirect supervisor *and* you disclose the relationship.
>
> If you can't provide a reference for a work experience (e.g., because you left on very bad terms), be prepared to honestly discuss that situation in an interview.
>
> Customize the list of references you use based on the type of job for which you are applying.

Reference checking is an important part of the hiring process, and I have seen many ostensibly ideal candidates rejected because of bad reference reports.

Many people put the "References" header in the résumé and then write "Available upon request." But I think this is a bad idea as it wastes valuable résumé space. You should always list your references if they are requested, and you should have your list readily available if it is not explicitly requested.

5.2.18 Order Matters

The order in which you list the elements in your résumé matters. A recruiter or hiring manager is going to be sifting through perhaps hundreds of résumés. She won't be able to read each résumé in detail, but a well-written first section might cause her to pay more attention to yours.

I believe that the best résumés start with either your experience or your education—whichever is more impressive. If you have little work experience, make your education the starting point of your résumé. If you have extensive work experience, but no college degree, lead with your experience. Leading a résumé with a list of hardware and software skills is rare, but I have seen that format used by software engineers and computer scientists, particularly when both the education and professional experience areas are weak.

Use good judgment in ordering the elements of your résumé and follow the conventions of your professional discipline.

5.2.19 Things to Avoid on a Résumé

Note that I have not recommended a particular font or format for a résumé. Use good judgment here. Whatever format you choose, avoid using any gimmicks, including multiple colored fonts, "unusual" fonts, strange formatting, iconography or embedded images, weird paper, and perfume. Sophisticated employers and headhunters use software that does keyword searches and other forms of intelligent analysis of your résumé. Unnecessary embellishments can interfere with that process, thus hurting your prospects. In addition, using these gimmicks can bring negative attention to your résumé.

I advise against embedding your picture in the résumé. You don't want to be hired or rejected based on your features, dress, perceived ethnicity, or anything else that can be inferred from a picture. For every person who might be impressed by your appearance, someone else might be unimpressed.

Strictly avoid humor in your résumé. While some people might get the joke you so cleverly embedded in, say, the job experience section, others may see your wit as offensive or awkward. If you are a funny person, and it seems appropriate, save your sense of humor for the interview. I have seen résumés and cover letters where the candidate claims that they have a good sense of humor, then at the interview the candidate appears to be humorless. It is better not to claim any particular personality attribute, but exhibit that attribute at the interview. Also, being funny usually isn't a job requirement unless you hope to be hired as a comedian.

Finally, delivering your résumé via special means (e.g., by private courier) doesn't usually help—many hiring managers see that as pushy and ostentatious. Let the quality of your experience and education win you the job. Getting the job via elaborate self-promotion will only lead to disappointment later if you can't deliver on inflated expectations.

When you have drafted your résumé, have several unbiased persons look at it and give you their honest impressions. Don't become defensive about the things they don't like: simply adjust the résumé accordingly.

5.2.20 Honesty Is the Best Policy

There are programs that parse résumés and extract certain keywords that might cause your résumé to bubble to the top. Applicants that know about these programs or are simply willing to exaggerate their skills will often include or emphasize proficiencies that they do not really possess.

You must be absolutely honest in your assessment of your abilities. Employers will likely test your claims, and Google can find postings that you may have made to blogs, user groups, Facebook, Twitter, and so forth. These postings might reveal your ignorance, a contradiction in claimed facts, or even embarrassing comments and pictures that might immediately disqualify you from further consideration.

Prospective employers expect applicants to have real expertise in one or two areas. Don't be shy about noting these. However, no one is an expert in everything, and claims of expertise in too many areas raise questions about your credibility. Again, honesty is the best policy.

5.2.21 Examples

Look at my résumé.

In the academic world, a résumé is called a Curriculum Vitae or "CV." My CV fits on one page and includes most of my work experience. I have a separate publications list (which is twenty-eight pages long), a long list of consulting clients and engagements, and a list of references, but I'll spare you those. I used to keep a list of hardware and software, but I tossed that long ago. I compressed my early career experiences into one item ("Various software engineering, consulting, and software management positions"). I don't elaborate upon these because I am no longer looking for a position in the industry. If one day I change my mind, I would significantly expand on these early experiences.

My résumé probably looks very different from yours because I am an educator, not a practicing engineer, and because I use my résumé for different purposes. I am not looking for a job, but I include my CV when applying for grants, and I send it to prospective consulting clients.

PHILLIP A. LAPLANTE, CSDP, PE, PhD
1313 Mockingbird Lane
Los Angeles, CA 9002
(211) 849-1111
plaplante@psu.edu

EXPERIENCE

Professor of Software and Systems Engineering *Pennsylvania State University, Great Valley Graduate Center, Malvern, PA, (Associate September 2001—2005, Full 2006—present).* Conducted graduate level scholarly research, teaching, and advising. Highlights include publishing 35 books and 250 papers in refereed periodicals, winning more than $4 million in grants, and developing six new courses.

Chief Technology Officer *Eastern Technology Council, Wayne, PA, January 2002—July 2009.* Served as technology consultant for regional nonprofit agency providing "contacts and capital" to more than 800 member companies. Created and led regional community of practice, The CIO Institute, with more than 80 member CIOs.

President *Pennsylvania Institute of Technology Media, Pennsylvania. June 1998— September 2001.* Chief executive officer of a private two-year technical college. Increased enrollments by 30% while reducing operating budget deficit by 50% per year over a three-year period.

Dean, Division of Science, Math and Technology *Technology and Engineering Center of New Jersey Institute of Technology/Burlington County College, Mt. Laurel, NJ. February 1995—May 1998.* Founding dean of a unique, joint venture that delivered over 40 degree programs (associate through doctorate) in science, mathematics, engineering and science.

Department Chair (from June 1992), Associate Professor of Computer Science *Department of Math/CS/Physics, Fairleigh Dickinson University, Madison NJ, September 1989—January 1995.* Served as Department Chair from 1992 to 1995. Chair responsibilities included supervision of 10 full-time and 14 part-time faculty members and staff, administration of a $600,000 budget, curriculum oversight and liaison with central administration.

Various Software Engineering, Consulting, and Software Management Positions *1982—1989.* Includes work at AT&T Bell Laboratories, NASA, UPS, Charles Stark Draper Laboratories, Canadian Defense Forces, and Singer-Kearfott Navigation Systems.

EDUCATION

- **PhD Computer Science and Electrical Engineering,** *Stevens Institute of Technology,* Thesis Area: Image Processing, 1990.
- **M.B.A.** *University of Colorado at Colorado Springs,* Concentration in Organizational Leadership, 1999.
- **M.Eng. Electrical Engineering,** *Stevens Institute of Technology,* 1986. Specialization in Digital Signal Processing.
- **B.S. Systems Planning and Management,** *Stevens Institute of Technology,* 1983. Concentration in Computer Science.

LICENSES AND CERTIFICATION

- Registered Professional Engineer (PE) Commonwealth of Pennsylvania
- Certified Software Development Professional (CSDP)
- Fellow of IEEE and SPIE

Now let's look at some résumés that are probably more similar to yours. First, consider the following job advertisement for a Software Engineering position:

> Teleteque, a major telecommunications firm is seeking a senior software engineer to provide full-lifecycle support for Teleteque-embedded telecom products, with an emphasis on VoIP applications Duties include creating and reviewing software design documents, writing unit test plans and code, leading testing activities, and software release.
>
> The ideal candidate will have 3+ years significant experience with Linux, Unix or Solaris; 3+ years experience in VoIP call processing and signaling protocols, including SIP/H.248/Megaco/MGCP/IMS; 5+ years solid experience with C++, including STL and templates; and a strong background in object-oriented analysis and design for engineering applications. Real-time embedded C++ software development a plus. Please send your résumé, along with a cover letter and list of references, via email to Ted Francis at tfrancis@teleteque.com.

Now consider this simplified and shortened résumé for John Proudface:

John G. Proudface
14 Fishing Pole Drive
Orlando, FL 32801
(407) 414-2309
jproudface@gmail.com

Summary: A seasoned professional with diverse software development experience in several environments and languages and extensive, real-time embedded systems programming experience. Seeking to become a key contributor to team building telecommunication applications.

Senior Software Engineer, Roadkill Systems, Orlando, FL, December 2008 to Present

Lead engineer for the JackRabbit voice over IP speech-to-text client. Led a team of three to develop the software requirements, implementing the system using Gang of Four patterns and Meyers' design-by-contract objected-oriented design principles, wrote and tested C++ code, and provided ongoing maintenance and customer support.

Software Engineer, Roadkill Systems, Orlando, FL, June 2004 to December 2008

Developed C++ and C code for a family of handheld device applications based on specifications provided by design engineers. Employed best practices for embedded real-time software debugging and testing of these applications. Created scripts and tested a variety of VoIP products in numerous configurations, including PCs running versions of Windows and Linux, and workstations running several Unix variants. Maintained rigorous test records and reported test results to development team and managers.

Education: BS Electrical Engineering, May 2004, Pennsylvania State University.

Now consider the résumé for Roger Doofus, who has identical work experience to Proudface.

Roger B. Doofus
2167 Swordfish Lane
Orlando, FL 32801
(407) 818-0919
r.b.doofus@sametech.com

Summary: More than three years significant experience with Linux and Unix, extensive experience in VoIP call processing and signaling protocols, including SIP/H.248/Megaco/MGCP/IMS. Significant experience with C++ and real-time embedded C++ software development. Seeking position where I can use these vast skills to help my employer.

Senior Software Engineer (since Dec 2008), Roadkill Systems, Orlando, FL, June 2004 to Present.
Responsible for all aspects of a family of voice over IP speech-to text-clients. Employed solid object-oriented techniques throughout life cycle in C++ and C code. Worked on real-time embedded code in C/C++ for held device applications. Tested a wide range of VoIP products in numerous configurations.

Education:
BS Electrical Engineering, 2004, Pennsylvania State University. GPA 3.9/4.0

Interests:
Fly fishing, rugby, white-hat hacking

Look at the difference between these two résumés intended for the same job. Both are based on nearly the same work experiences. Proudface is much more subtle in tuning his résumé to fit the job advertisement. Proudface uses action verbs and details how he used object-oriented software design techniques. He also breaks up his work experience at Roadkill to show how he was promoted from Software Engineer to Senior Software Engineer in less than five years.

Doofus appears to have copied and pasted the job advertisement into his résumé. Doofus parrots claims about what he did in his work, but there is no evidence that he actually knows what he is talking about and he collapses all his work experience into one nebulous entry. Doofus is using his work e-mail—this is inappropriate and risky if Doofus' company monitors employee e-mails. Doofus also included his GPA and lists his hobbies (and "white-hat hacking" is not a desirable one), which I discourage for the reasons mentioned previously.

5.3 Transmittal Letters

Transmittal or cover letters accompany another artifact, such as a résumé, proposal, or a defective product return. Sometimes transmittal letters are an unaccompanied request for information. The basic goal of any transmittal letter is to briefly and concisely state your purpose.

Following is an appropriate cover letter for John Proudface to include with his résumé in response to the Teleteque job opportunity. Because the job requested that résumés be sent via e-mail, this cover letter is actually the body of an e-mail to Mr. Francis:

> Dear Mr. Francis:
>
> Please find attached my résumé in response to your job advertisement in the December 10th *Clearwater Times* Jobs section.
>
> I believe that my experience and education are precisely in line with your requirements. I hope that I will have an opportunity to highlight my qualifications in a personal interview.
>
> Thank you for your consideration.
>
> Sincerely,
> John Proudface

Note that there are no embellishments here, just a short, simple letter of introduction.

You can also use a transmittal letter to highlight any special coincidences that might exist. For example, if you are applying for a job and are acquainted with some key employee, you might note that fact. Avoid any gimmicks such as jokes, strange formatting, or inappropriate references.

5.4 Writing Letters of Reference

You may be asked to write a letter of reference for a friend or colleague, and you surely will be asking others to write one for you. Therefore, it is important to discuss the writing of an effective letter of reference.

The standard elements of a letter of reference are as follows:

Explain who you are and why you are writing

Explain how you know the candidate and for how long

Describe the candidate's strengths and weaknesses

Comment on the candidate's suitability for the job in question

Summarize your recommendation

Customized letters are much more powerful than generic ones. "Dear Mr. Jones" is more personal than "To Whom It May Concern." "I am writing to support Mr. Francis for the position of Director of Engineering" is more effective than "I know Mr. Francis is the right person for the position for which he has applied."

The most important aspect of a letter of reference is what is not written. If you cannot say something positive about a candidate's punctuality, don't mention it. Equivalently, when you read a letter of reference, look for the absence of positive comments about specific aspects of the candidate—demeanor, punctuality, work quality, honesty, reliability, and so on. Anything less than a positively glowing letter of reference means that the writer is equivocating and you should probably avoid that candidate.

Following sections provide examples of disguised (but real) letters of reference that I have written over the years. A template for a general letter of reference can also be found in Appendix B.

5.4.1 Letter of Reference for a Subordinate

The first is a recommendation for a person who worked for me when I was a college dean:

> Dr. Ralph Kramden
> Chair
> Department of Textiles Merchandising & Interiors
> The University of Flatbush
> Brooklyn, NY 11226
>
> Dear Dr. Kramden:
>
> I am writing in support of Ms. Trixie Norton, who I understand has applied for the position of lecturer of fashion merchandising in your department. In short, I give her my highest recommendation without reservation.
>
> To give you a context for my recommendation, please let me tell you briefly about myself. I am currently a Professor of Software Engineering at Penn State University. I returned to research and teaching after a successful career as an academic administrator (chair, dean, and president of a small two-year college). As dean at Brooklyn County College, I was Ms. Norton's direct supervisor. In fact, I hired her after an extensive search and supervised her for more than two years.
>
> Trixie created a vibrant and successful fashion design program from scratch. She designed the labs, wrote the curriculum, taught many of

the courses, and supervised adjunct faculty. In all of these things she was very successful. Trixie was also deeply involved in marketing and promoting her program and other programs in the division, such as the computer graphics and multimedia program, where she made many contributions.

Her work with the students was outstanding. She was always enthusiastic, energetic, patient, and kind. Because of these attributes, she was beloved by the students and she was a joy to supervise. She took direction very well, provided feedback, was diligent and hardworking, and was a positive role model for my other directors. I would most certainly hire Trixie again if the opportunity arose.

In closing, I don't think you will find a more knowledgeable, personable, and hard-working person for the lecturer position. I hope you will be lucky enough to hire her. I therefore give her my highest recommendation without any reservation whatsoever. Please do not hesitate to contact me if you have any questions.

Sincerely,
Phillip A. Laplante, CSDP, PE, PhD

This is an unequivocal recommendation. I knew "Trixie" well and could attest to all of her positive attributes. The key phrase is "I give her (him) my highest recommendation without any reservation whatsoever." I would not include this sentence in a letter if I did not believe it to be true.

5.4.2 Letter of Reference for a Casual Acquaintance

The next letter is for someone I did not know very well. You can see that although the letter is positive, it is somewhat evasive. I really can only comment on what I knew from his résumé, his overall reputation, and a few brief exchanges over the years. But I did feel that I knew him well enough to give him an endorsement. If someone were to ask me to write a recommendation letter, and I felt that I did not know that applicant well enough, then I would decline.

Dear Review Committee:

I am writing in support of Dr. Roger Wendell, who I understand has applied for a position as a senior researcher at MMGood Systems. I have known Dr. Wendell for 12 years since meeting him at a systems integration conference in Paris. Since then, we have been in regular communication and have continued to collaborate on the program committees of various conferences on real-time image processing, and as members of the editorial board of the journal, *Real-Time Imaging*.

Dr. Wendell has excellent academic experience having been a member of the faculty at several research universities. His research during these periods has been outstanding, having resulted in a number of publications and presentations in top venues.

While Dr. Wendell is an excellent researcher, he is also blessed with that rare combination of industrial and practical experience, coupled with a solid theoretical foundation. He has broad international experience and has worked on a wide range of imaging, digital communications, and signal processing projects. For several years now he has built and managed a successful consulting firm. Undoubtedly, his management skills are as excellent as his technical abilities.

Dr. Wendell is also a most pleasant individual. He is friendly, very hard working, and has outstanding communication skills. He will be an excellent mentor to your junior staff and a wonderful colleague to his peers, staff, and supervisors.

I therefore give Dr. Wendell my highest level of recommendation with great confidence that you will be most satisfied with his work. Please do not hesitate to contact me if you have any further questions.

Sincerely,
Phillip A. Laplante, CSDP, PE, PhD

5.4.3 Generic Letter of Reference

The final recommendation letter pertains to someone who was applying to a graduate program in primary education. This letter is somewhat generic because I was not given a specific addressee, and I did not know much about the applicant's academic performance.

To Whom It May Concern:

I am writing in support of Ms. Cheryl Mays, who has applied for a position with your organization.

To give you the context for my recommendation, I note that I am a friend of the family and have known Cheryl all of her life. While it is difficult to suppress my personal adoration and respect, I believe I can objectively evaluate her.

I am sure it is clear from her academic record that she is smart and energetic as she did extremely well in a difficult major, at an excellent college. It bears noting, however, that her expansive service activities were borne from true social responsibility and not from any mandates or peer pressure. Her travel abroad and work with those from other cultures further evidence rare interpersonal skills. I think that in our current environment these character attributes are more valuable than an obituary-like employment history.

Cheryl's personality is one of healthy balance. She is thoughtful yet responsive, respectful yet playful, serious yet cheerful. Her varied experiences demonstrate true qualities of leadership, creativity, scholarship, and pedagogy. These activities, in and outside of college, illustrate what I have known for many years—that she is a quiet leader who is comfortable in both group settings and in one-on-one interactions.

In closing, I note that I have known many young adults through 20 years of teaching and community service. I believe that Cheryl ranks in the top one percent of those in terms of intelligence, diligence, and compassion. She will bring energy, enthusiasm, broad knowledge, and character to your organization. I therefore give her my highest recommendation without reservation.

Please do not hesitate to contact me if you have any questions.

Sincerely,
Phillip A. Laplante, CSDP, PE, PhD

Generic letters for individuals can be hard to write because you must find positive attributes without being too vague. In this case, I did not know Cheryl's academic background very well, but I knew her all her life and I could positively vouch for her personal qualities.

5.4.4 Form-Based Letter of Reference

In some cases, the letter of reference is not a letter at all but an electronic fill-in-the-blanks form. Here is an example of a modified electronic reference form for a large company:

How long did you work with this candidate and in what capacity?

The applicant is being considered as a _____. Do you think he/she is qualified?

What were his/her major accomplishments?

How did his/her performance compare with that of others?

Did he/she meet all of your expectations for this position?

How did this person get along with others?

What kind of job do you think he/she is best-suited to do?

How would you rate his/her ability to handle stress?

Why did he/she leave your company?

Would you rehire this person? Why or why not?

If you could give one piece of advice to his/her future manager, what would that advice be?

Do you have any other comments regarding this individual that would be helpful in making the hiring decision?

Although it may be more challenging to provide the same level of detail in the responses to these questions, the principles for completing electronic references are essentially the same as for writing letters of reference.

5.5 Memos

You will need to write many memos as a technical professional. Good memos are purposeful and persuasive.

The format of a memo is quite informal but depends greatly on the local conventions of your workplace. The basic elements of a memo are

To: Name of addressee/recipient

From: Who wrote the memo

Date: The date the memo is sent

Subject: A memorable handle for the subject

State your purpose

Elaborate on your purpose

Summarize the memo and state any actions that you will be taking

The following example is a disguised version of a memo that I wrote my dean when I was a junior faculty member and looking to impress her with my ambitions:

To: Dean Jones
From: Phil Laplante
Date: March 19, 1992
Subject: Cinematic Series Proposal

I would like to propose a weekly film series for next fall semester. The goal is to examine the evolution of the impact of computers as portrayed in film.

To encourage participation, I think the screenings should be open to the University community for free. It might also be nice to provide beverages and snacks. Finally, the series could culminate in an "all-night" film festival where several of the films could be shown. After each film, a discussion session will be conducted.

The following is a tentative film list:

Film Title	Production Company	Year Released
Forbidden Planet	MGM	1956
Desk Set	TCF	1957
2001: A Space Odyssey	MGM	1968
Colossus: The Forbin Project	Universal	1969
Westworld	MGM	1973
Tron	Walt Disney	1982
War Games	MGM-UA	1983
The Terminator	Orion	1984
Terminator II	Carolco/TriStar	1991
Lawnmower Man	Columbia/TriStar	1992

In addition to these, I would like to show several relevant episodes from the old *Star Trek* and *Twilight Zone* series. I have already had a preliminary discussion with the dean of libraries, who seems enthusiastic about this project. She feels that the logistical problems, such as obtaining the films, venue, and so forth, can be easily resolved.

However, I ask your help, should you approve this project, in soliciting the participation from other academic departments, public relations, student affairs, and so forth. I will encourage the participation of the ACM Club, which I advise.

After receiving your approval, I will prepare a budget, final film schedule, and begin any other preparations necessary. I will be requesting some funding to support the activity.

I am excited about this idea! In addition to being a genuine intellectual activity, it provides needed social events for the students and good publicity for the college, our department, and the ACM club. I hope that you agree.

This memo seems somewhat simplistic to me now, but it displayed my raw enthusiasm at the time. My dean agreed to the proposal for the film series, I believe, because it was a good idea and my memo was persuasive.

5.6 Meetings, Agendas, and Minutes

5.6.1 Meeting Invitations

An invitation to participate in some activity is another common form of written letter for technical professionals. An effective invitation letter describes the purpose of the activity, the participant's role, and where and when the meeting will be held. The letter should also discuss logistics such as travel and transportation cost reimbursement (if any). There should also be a call to action and a deadline by which a response is requested or required.

For example, here is an invitation to a meeting regarding the software engineering licensure project:

> Dear Mr. White:
>
> You recently expressed interest in participating in the effort to develop a professional engineer licensing examination for software engineers in the United States, and I hope that you are still willing and able to participate in this effort.
>
> I am the chair of the Software Engineering Licensure Exam Development Committee. On behalf of the Software Licensure Project Steering Committee, I am pleased to invite you to attend a meeting on September 30—October 1 at Penn State's Great Valley Campus (near Philadelphia in Malvern, PA). During this meeting, we will commence examination development with an important first step—the creation of a Professional Activities, Knowledge, and Skills (PAKS) survey. This survey will be administered to a pilot group and later to a broader group of practicing software engineers. The survey results will then be used to define the exam content areas and the proportion of questions for each area. The exam development process moving forward will be explained in more detail at the meeting.
>
> Reasonable travel expenses will be reimbursed and all of your meals will be covered. I will send you more travel and expense details soon, but at this time please indicate your willingness to participate and your availability for the meeting.
>
> You have been selected for this meeting because of your knowledge and experience from amongst many other applicants. I hope you can serve in this important capacity, but in any case, please respond to me by August 1.
>
> Thank you very much,
> Phil
>
> Phillip A. Laplante, CSDP, PE, PhD
> Chair, Software Engineering Licensure Exam Development Committee

This letter contains enough information for an invitee to understand the motivation for the project, logistical details, and what is expected of participants.

5.6.2 Agendas

You will have to prepare agendas for meetings many times in your career. A successful meeting starts with a focused and well-organized agenda. The elements of an agenda include:

Meeting name

Purpose of the meeting

Location

Start time and length

List of topics to be discussed with lead person and time allotted identified

Proposed dates for follow-up meetings

Here is an example of a simple agenda for a meeting of a community of practice for CIO-level executives:

AGENDA
IT Directors Advisory Board
June 5, 2017

Welcome and introductions	(Phil)	(12:00–12:05)
Review of this year's program	(Amy)	(12:05–12:25)
Discussion of upcoming events this year	(Phil)	(12:25–1:15)
July meeting(s)?		
Best times/days for other events		
Other items to discuss possible institute sponsorship	(All)	(1:15–1:25)
Wrap-up and action items	(Phil)	(1:25–1:30)
Adjournment		(1:30)
Next meeting: October 15, 2018		

Notice that I have identified the lead person for each agenda item and the allotted time for discussion. The next meeting date is also given. Appendix B contains a template for an agenda.

5.6.3 Meeting Minutes

After the meeting, someone should prepare a record of the meeting (called the meeting "minutes"). Characteristics of well-written meeting minutes include completeness, clarity, and a list of action items for further follow-up. A list of attendees and absences should also be included. Ideally, minutes convey to those who were not at the meeting the essence of what was discussed and decided. The minutes may be written in an informal manner. You can begin writing meeting minutes using the meeting's agenda as an outline.

Here is a set of minutes for the IT Directors meeting agenda just shown:

Minutes
IT Directors Advisory Board
June 5, 2017, 12:00

Attendees
Phil Laplante, Amy Nash, Arthur Sprague, Rodney Taylor, Susan Felder

Welcome and introductions
After general welcome and introductions, the events of last year were discussed.

Review of this year's program
Some discussion ensued over whether the events should be held in the morning or at lunchtime, and whether they should be held in the city or in Chester County or elsewhere. The consensus was that a mix of times and venues was still most desirable. A New Jersey event (e.g., Camden Aquarium) was also suggested.

Discussion of upcoming events this year
It was agreed that the July 6 "hot topics" roundtable should be removed because of expected low attendance.

Desired topics for the rest of the year included:

    IT/Business alignment
    Cultivating IT leaders
    An event on the Philly wireless initiative
    Business intelligence

Other items to discuss
Amy mentioned that there have been some discussions with a potential overall sponsor for the CIO Institute. In general, we are looking for such a sponsor and Amy will be keeping the board updated.
A discussion was conducted on the desirability and feasibility of some kind of CIO retreat in a location such as Bermuda. It was concluded that such an event would be welcomed by the membership.

Action items
Amy will investigate the possibility of sponsorship for a CIO retreat.

Adjournment
The meeting adjourned at 1:45.

You should send a draft of the meeting minutes to all attendees before general distribution. This courtesy allows for technical corrections to be made and embarrassing misquotes or misinformation to be removed. A template for meeting minutes can be found in Appendix B.

5.7 Customer Relations Writing

Throughout your career you will write all kinds of paper and e-mail correspondence, both technical and nontechnical. You will write to current and future customers, vendors, government agencies, and colleagues. Writing to customers is a very special case because of the financial and legal ramifications of any misunderstanding. You must also be exceptionally polite in such correspondence.

5.7.1 Vignette: A Customer Inquiry Letter

I wrote my first book on experiences I had running my own small computer support business [Laplante 1992]. I'll tell you more about this book in Chapter 8. As an extended case study, consider the following disguised letter from a reader of my book, that is, a customer letter that I received right after the book's publication.[1]

Dear Dr. Laplante:

My name is Gus Brady and I am seeking advice. As I secured your name from the writings of your recent publication, I hope you may provide me helpful insight.

I am considering a substantive career move, one founded on a complete departure from my professional and academic upbringing. Briefly, I am forty-three years old and I have been working in the banking and real estate industry for twenty years. I have an undergraduate degree in business. I am somewhat tired of working in a traditional office environment and for other people. I wish to start my own business and refocus my attention in a technical capacity. In my opinion, the support service of the computer industry is a sector of great possibility and growth. Not only do I want to be a part of the business side of this opportunity, I wish to experience it literally "hands-on." My thoughts are to return to school for a two-year technical degree and subsequently work through an apprentice program. Within four years, I hope to open my own shop (hardware support) and ultimately remove myself from the day-to-day maintenance. My duties would gravitate to business planning and marketing.

With this description, I would greatly appreciate your advice on the viability of my plan, especially given my age. Your insight is also valued regarding the industry outlook and the related profit margins. Finally, and most importantly, an indication of technical schools that you could recommend would be most appreciated.

I realize that your time is quite valuable and for this reason a response through any medium you prefer would be most appreciated. A self-addressed return envelope is enclosed; my direct line at work with voice mail, my fax number, and my home phone number can be found on my enclosed résumé.

My intent is to make an informed decision and your insight would assist me greatly in moving in the right direction. If you have suggestions of alternative industry(ies), such as digital or analog support, I would very much like to hear them. Thank you again for your time.

Sincerely,
Gus Brady

This letter is an effective transmittal letter. Although the letter is not brief, it is very honest, and a reader clearly can understand the intent of the writer.

5.7.2 Vignette: Response to a Customer Inquiry Letter

The inquiry letter shown in Section 5.7.1 was so intriguing that I felt compelled to respond promptly. Here is my response, with identifying information disguised:

Dear Gus:

Thank you for your nice letter of June 14. I am flattered that you value my opinion. I am happy to share my insights with you. First, let me point out that I have not been active in the computer service industry for the last three years. As this technology changes very rapidly, I am not even sure I am qualified to correctly respond to your inquires. With that disclaimer, I will offer my opinion.

When I left the business in 1990, the trend in computer service was away from small shops to large companies. It was very difficult for a one-man operation to stock all the parts necessary, and to be expert in the many different facets of computer service. I must say, though, that I was making good money on service calls (but losing money on hardware, as I could not compete with the mail order businesses). The problem with any business dealing in computer hardware is that most people are semi-computer-literate (they think they know more than they do), and they all have a catalog in hand. These people wanted me to sell hardware at catalog prices, but provide 24-hour, customized service![2]

Don't let me discourage you, however. I recently saw a relatively small computer service company that had a very interesting strategy. The company guarantees an eight-hour response time for a service call (this is a must) and charged a flat $145 fee for the call. This fee was the same whether a plug was loose or a disk drive needed to be replaced. I think this is the way to go. It eliminates nuisance calls, while at the same time you don't need to compete with catalogs—i.e., you don't need to reveal how much you charge for hardware. The difficulty here is finding a flat fee that will result in a profit, but which a customer will buy. One way to do this would be to list the top 100 service call scenarios, affix a cost to each, and then calculate the expected cost of the service call based on the probabilities for each of the 100 scenarios. You can see how to do this by reading the section on "expected value" in any probability book. What you charge for each service call should be significantly more (about twice, in my opinion) than the expected cost of the call.

As far as learning how to repair PCs, you can take courses at local community colleges, specialized schools like Spangler Institute, or through commercial education companies like Blue Bird Education. But there is no substitute for experience. I used to learn how to repair PCs using customer machines as guinea pigs. I was lucky, and never lost a patient.

Anyway, I wish you luck in your new venture. Depending on your local market conditions, your business skill, and your eventual skill in repairing PCs, I think you can make a lot of money doing this. But please don't let me know if you become a millionaire—I will be jealous and my wife will kill me.

Best regards,
Dr. Phillip A. Laplante

Gus received a nice response from me and free consulting advice because he wrote such an enticing letter. Notice how I used a very informal tone and humor, which was appropriate in this case.

5.8 Press Releases

It is possible that you will be asked to write or contribute to a press release. A press release is a short statement that is sent to local newspapers and online news services to announce some particularly meaningful event. Press releases must be effective because newspapers and online services don't give away valuable print space. The central point of the press release must be "newsworthy," concise, and important. Press releases ought to be neutral in tone so that they may be the basis of news stories. Press releases that sound like advertising will be disregarded.

The elements of a press release are as follows:

A persuasive title

A concise description of the news

An official point of contact for more information

The press release is distributed to an official list of contacts (e.g., reporters) that your company's market or public relations department keeps.

Here is an example of a press release I cowrote[3] for the software engineering licensure project:

> IEEE-USA to Serve as Lead Technical Organization in Software Engineering Professional Licensure
>
> (May 10, 2010)
> IEEE-USA and the National Council of Examiners for Engineering and Surveying (NCEES) finalized an agreement to jointly develop the principles and practices of software engineering examination, unofficially known as the "PE examination for software engineers." The exam completes the set of tests needed to license professional software engineers in the United States.
>
> IEEE-USA is working closely with the IEEE Computer Society, the National Society of Professional Engineers, and the Texas Board of Professional Engineers to provide financing and technical subject matter experts for the effort. This project came to fruition after a majority of IEEE members working in the field of software engineering surveyed indicated that a path to professional licensure was needed. The desire of IEEE members, coupled with the request from ten state boards of engineering licensure, provided the motivation to begin the process.
>
> NCEES requires that any new licensing examination discipline have a technical society as an official sponsor. Because IEEE is the leading professional organization for software engineering professionals, the board of directors approved a proposal for IEEE-USA to serve in this capacity in November 2009. The contract was drafted and signed between IEEE-USA and NCEES in June 2010.
>
> Dr. Phillip A. Laplante, a licensed professional engineer, IEEE Fellow, and professor of software engineering at Penn State University was selected to chair the examination development committee. Other members of the committee are being recruited and it is anticipated that the PE examination will be available for state boards to administer in mid-2012.
>
> For more information, please contact: Phil Laplante at plaplante@psu.edu.

Normally, phone and fax numbers are also provided with the contact information.

5.9 Presentations

As a technical professional, you will be called upon to present your ideas, whether they are a proposal for a new project, a progress report for an ongoing project, or a post-delivery review of a completed project. A technical presentation is a form of technical writing that is tightly coupled to a talk by the presenter. While it is out of the scope of this book to discuss the best practices of public speaking, it is worthwhile to describe how to put together effective visual presentations using commonly available software tools.

The best presentations are punchy and use minimal text and more graphics. The presentations should not be a copy of the speech to be given. The slides should enrich and complement the talk with tables, exhibits, and pictures illustrating the talk's main points.

Slides based on bullet points may be used to guide the audience through the main points of the presentation, keeping the audience in context as the speaker provides more detailed supporting discussion for each point. The speaker may also have more detailed notes, to help remember items and details to mention during the talk. The finer level of detail in the speaker's notes may be omitted from the slides. The following example shows the process of evolving the detailed speaker's notes into summary headings that act as a contextual map throughout the presentation.

5.9.1 Vignette: A Presentation on Cyberpandemics

Here is an example from a presentation on "cyberpandemics" that I gave to the Greater Philadelphia Chamber of Commerce in 2010. A cyberpandemic is a worldwide cyberattack that triggers devastating second-order effects, such as massive power disruption, disabling of a water supply, or knocking out communications [Laplante, Michael, and Voas 2009]. I was asked to give a forty-five-minute presentation, allowing fifteen minutes more for questions. In my experience, I speak for about three minutes per slide. So, my target for this presentation was about fifteen slides. You may find that your slide-to-talk-time ratio is different—keep track of this metric as you give presentations because this will help you gauge the right number of slides to prepare.

For the cyberpandemics presentation, I started out with a set of about twenty verbose slides. Two slides from the presentation are shown in Figures 5.1a and 5.1b.

Having as many words on a slide as I had included on the second slide of Figure 5.1 makes the slide unattractive to the audience. Realizing that these slides were no more than a "script," I revised them to simple headings, as shown in Figures 5.2a and 5.2b.

Conditions for a Cyberpandemic

Complexity

    people packed too tightly in cyberspace

    complex social interactions (difficult to track)

Multiple simultaneous attacks

    multiple, orthogonal vector mechanisms

    one or more non-cyber components

    analogy is attack to immune system and nervous system

Symptoms emerge long after infection has occurred

    makes widespread dispersal likely, difficult to prevent

Infected elements in the cyberattack may be coordinated to work in concert (e.g., Botnet worm)

(a)

Cyberpandemics—Who Are the Threats?

Rogue governments

Rogue elements of legitimate governments

Terrorist groups

Corporations and consortia that may profit

Malicious pranksters (various motivations)

An individual?

Rarely (never?) by accident—experiment gone wrong

(b)

FIGURE 5.1
Two verbose slides from a presentation on cyberpandemics. (a) Verbose slide for discussing necessary conditions for a cyberpandemic attack. (b) Verbose slide for discussing potential initiators of a cyberpandemic attack.

Necessary Conditions

Complexity

Multiple simultaneous attacks

Symptoms emerge long after infection has occurred

Infected elements may be coordinated

(a)

Cyberpandemics: Who?

Rogue governments

Rogue elements

Terrorist groups

Corporations and consortia

Malicious pranksters

Individuals?

By accident?

(b)

FIGURE 5.2
Two improved slides from a presentation on cyberpandemics. (a) More succinct version of slide shown in Figure 5.1a. (b) More succinct version of slide shown in Figure 5.1b.

Finally, I decided to winnow the text further and add some copyright-free clip art to add interest and to help guide the story as I made my presentation. The final slides are shown in Figures 5.3a and 5.3b.

The perfected slides are less verbose, have some color and imagery to hold the audience's attention, and provide some visual cues about my path through the material. I was also able to combine some slides, so that the final slide count was seventeen, which fit the time allotted. The multistep revision process I used to obtain the final slide set is similar to the revision process used in other forms of writing.

Necessary Conditions

Complexity

Coordinated Infections

Late Emerging Symptoms

Multiple
Simultaneous Attacks

(a)

Cyber Pandemics : Who?

Rogue Governments

Terrorist Groups

Corporations and Consortia

Malicious Pranksters

(b)

FIGURE 5.3
Perfected slides from a presentation on cyberpandemics. (a) Pictorial version of slide shown in Figure 5.1a. (b) Pictorial version of slide shown in Figure 5.1b.

5.10 Marketing and Sales Materials

As a technical professional, you may be asked to participate in writing marketing and sales materials. These materials include sales brochures, product specification sheets, and advertising copy. Or, you may wish to market your own services as an independent consultant, in which case you will have to write your own marketing and sales materials. In either case, I'd like to provide you with some advice.

Writing marketing and sales materials is quite different from writing straight technical material. In the latter case I have emphasized that you are trying not to evoke an emotional response, whereas in the former case you are. That is, with marketing and sales materials you are trying to convince a prospective buyer that your product is desirable and better than its competitive products in some way. Therefore, you may want to use some of the writing hooks that I have advised you to avoid such as humor, metaphor, and euphemism—these can be effective marketing and sales tools. You should avoid exaggeration and misrepresentation, however, even if you or your colleagues are tempted, for example, to advertise features that do not yet exist.

5.11 Exercises

5.1 Write a job description for your current job.

5.2 Write a job advertisement for your "dream job." Make the dream job realistic, that is, one for which you are qualified.

5.3 Prepare your résumé as if you are applying for your "dream job."

5.4 Write a cover letter to accompany your résumé in response to the advertisement you wrote in Exercise 5.2.

5.5 Prepare an invitation letter to your friends for a meeting to plan the thirtieth birthday party for a friend.

5.6 Prepare an agenda for the birthday party meeting noted in Exercise 5.5.

5.7 Prepare a press release announcing the thirtieth birthday of your friend.

5.8 Prepare a set of slides for a short (five-minute) presentation on a topic of your own choosing.

5.9 Prepare a (fictitious) agenda for the next meeting of an organization to which you belong.

5.10 Write a set of fictitious minutes for the meeting you discussed in exercise 5.9. Be humorous if you like.

5.11 For your favorite electronic product (for example, a television, smart home device or personal assistive device) write a 100-word advertisement.

Endnotes

1. This isn't the real letter—the writer of any letter holds the copyright, not the recipient. This is a fabrication in the spirit of the original letter.

2. In 1993 there were no competitive Web-based computer parts supply businesses. All the suppliers used paper catalogs.

3. Thanks to my friend Dr. Mitch Thornton, who cowrote this press release.

References

Laplante, P. A., *Easy PC Maintenance and Repair*, Windcrest/McGraw-Hill, Blue Ridge Summit, PA, 1992.

Laplante, P., Michael, B., and Voas, V., Cyberpandemics: History, inevitability, response, *IEEE Security & Privacy*, 7(1), 63–67, 2009.

SIL International, The ILR (FSI) proficiency scale, https://casemed.case.edu/registrar/pdfs/Scale_ILR.pdf, 1998. Accessed December 26, 2017.

6

Technical Reporting

6.1 Introduction

Technical reporting documents have much in common with their business counterparts but there are important differences. The former can be distinguished by a higher concentration of mathematics, data, formulae, and technical jargon; a need for greater precision; a higher attention to detail; and the likely inclusion of procedural descriptions. "Technical reporting" is also slightly different from the technical reports discussed in Chapter 4. The latter is a kind of archival writing, although informal, for mass consumption and reference. Technical reporting, conversely, refers to internal documents or reports for customers, agencies, or management that are considered confidential, proprietary, or classified.

As noted in Chapter 1, technical reporting includes the following classes of documents:

Progress reports

Feasibility studies

System and software design and architectural specifications

Proposals

Facilities descriptions

Manuals

Procedures

Planning documents

Environmental impact statements

Safety analysis reports

I would also include strategic plans and bug reports to this list. In Chapter 6, I discuss and provide examples for some of these document classes.

6.2 Technical Procedures

Technical procedures are descriptive documents that include step-by-step instructions for assembling, processing, or organizing things. A procedures document should contain the following (as applicable):

Purpose of the procedure

Safety information

Preconditions—what must be true before the procedure is conducted?

Qualifications of those who will conduct the procedure

List of materials

List of tools

Step-by-step instructions

Post conditions—what will be true if the procedure is followed correctly?

Troubleshooting section

List of frequently asked questions

The key to writing procedures manuals is to assume the reader is a novice, or to document your assumptions about the previous knowledge or experience needed. There is a special burden on the writer of technical procedures—even the smallest mistake can lead to catastrophe. Therefore, you need to have both experienced and naïve colleagues test your written procedures.

The wife of a former student recently published the delightful *Unofficial Harry Potter Cookbook*, in which she gives recipes for foods mentioned in the Harry Potter books [Bucholz 2010]. My student described to me the comprehensive recipe testing that his wife had conducted—preparing recipes repeatedly and varying the ingredients slightly each time to simulate a measurement error. She noted the sensitivity of the food to these errors and adjusted the recipes accordingly. The author showed great diligence and care in testing and revising her cookbook, setting a wonderful example of what you should do in writing technical procedures.

Here are a few excerpts of technical procedures that I have written.

6.2.1 Vignette: PC Repair Book

When I was finishing my PhD, I decided to make some extra money by starting a personal computer (PC) maintenance and repair business. After six months, I had depleted most of my savings because I was not a savvy businessman, but I did learn a great deal about PC maintenance and repair.

I converted this knowledge into a "how-to" book filled with technical procedures.

The following excerpt is the procedure for the low-level formatting of a hard drive, and is a good example of step-by-step technical instructions. Note that this procedure is outdated—the technology has changed significantly since 1995, so don't try reformatting your hard disk drive this way.[1]

There are two methods for low-level formatting a disk. Both methods are discussed below.

Static configuration

At the prompt
Are you dynamically configuring the drive? Answer Y/N
Select N. The next prompt will be:
Press "Y" to begin formatting drive C with interleave 3

Type y and press Enter. The program then asks you if you wish to format bad tracks (see subsequent section on this). There may be several other prompts warning you that all current data will be lost upon formatting; follow their instructions. The program displays the message ``Formatting'' while it formats the disk—a process that can take ten or more minutes depending on the size of the disk and the speed of the computer. The program displays a message when the format is complete.

After this step, proceed to the section called "Partitioning."

Dynamic configuration

If you type Y in response to the question "Are you dynamically configuring your drive," a screen similar to the following appears:
Key in disk characteristics as follows: ccc h rrr pp ee o
where:

ccc	=	total number of cylinders (1–4 digits)
h	=	number of heads (1–2 digits)
rrr	=	starting reduced write current cylinder (1–4 digits)
ppp	=	write precomp cylinder (1–4 digits)
ee	=	max correctable error burst length (1–4 digits) range (5–11 bits), default 11 bits
o	=	ccb option byte, step rate select, (1 hex digit), range = 0 to 7, default = 5 refer to controller and drive specification for step rates

The program needs these disk characteristics (which should be available from the manufacturer) to proceed with the low-level format. Unless your disk is larger than 80 Mb, the values for reduced write cylinder, write "precomp cylinder", max correctable error burst length, and ccb options are left blank; otherwise use the characteristics specified by the disk manufacturer. After you have entered these values you are prompted as follows:

> Are you virtually configuring the drive? Answer Y/N
> At this point the program is asking if you wish to do a virtual split. DOS version 2.0 allows drives of only 16 Mb, and DOS versions later than 2.0 up to and including 3.3 allow only 32 Mb. If you have a disk larger than 32 Mb, you will have to do a virtual split. If you are not doing a virtual split, respond by typing "N" and skip to the section on bad tracks.

This excerpt depicts the limitations of written procedures—a reader cannot ask clarifying questions. While you can try to anticipate frequently asked questions (and put those along with the answers in an appendix), it is impossible to anticipate every possible question. And if you add more variations to the procedures, the procedures become more complex. In writing procedures, you must balance completeness against clarity.

6.2.2 Vignette: Building an Aquarium

This example arose from a project to convert a computer monitor into a working aquarium. I was looking for a way to repurpose an old monitor into a fish tank, but could not find appropriate instructions. I developed an approach through trial-and-error and built a working aquarium. A friend saw my finished product and suggested that I build another for a charity auction. I agreed. In the process of constructing the aquarium, I took photographs and dictated my activities into a tape recorder as if I were hosting some kind of do-it-yourself show.

I converted the transcript of my ramblings and photographs into an article that was published in a hobbyist magazine [Laplante 2003]. The aquarium netted the charity $100. In addition, this was one of the few published articles for which I was paid—an unexpected pleasure. Here is a slightly revised excerpt from that article. First, I describe the list of materials:[2]

> In addition to a computer monitor, you will need only a few supplies that are easily found in local hardware or pet supply stores.
>
> *Fiberglass repair supplies, including fiberglass resin, hardener, and fiberglass mat or mesh.* Often, these items can be purchased as a complete kit in any hardware or automotive store. Be certain the items are recommended for marine use and that you purchase enough materials to cover approximately 9 square feet, although the amount needed will vary, depending on the size of your monitor. Nine square feet should be enough to seal a 17-inch monitor (the dimension of a monitor is measured diagonally across the screen, like a television).
>
> *Clear aquarium sealant.* The sealant must be clear, and it must be stated on the container that it is recommended for "marine or under waterline"

applications. Note that most clear silicon sealants found in hardware stores are not suitable for underwater marine applications or aquariums—in some cases, the sealant will actually poison the fish. Get this at a fish or pet store; the sealant is generally available in a squeeze tube—it should not be purchased in a caulking tube because this will make it difficult to apply to tight spots. Purchase at least 2 ounces.

A sheet of clear acrylic or plastic (often found under the trade name of Lucite). The size will depend on the monitor size. For a 17-inch monitor, you will need approximately a 3-by-3-foot sheet. Always buy a sheet that is larger than the screen opening. There should be protective paper covering both sides of the plastic; do not remove this paper until you're ready to install/ the window (more on this later). Do not use glass. It is difficult to cut to size and to seal. It is also less durable than plastic. Furthermore, we will be curving the window for a more authentic monitor look—something that cannot be done with glass.

In retrospect, this listing was too verbose. I tried to provide too much detail too soon. I should have listed the items, and then offered the details for each item. I was more concise in my description of the tools needed [Laplante 1995]:

> Only a few tools are needed for this project, and they include:
>
> Rubber gloves; you will need plenty of these—this is a messy job.
> Fine-tooth coping or keyhole saw or high-speed jigsaw (with fine teeth) or Dremel tool for cutting the tank top opening and the clear plastic
> Drill with ½-inch bit
> Phillips head screwdriver
> Paintbrushes
> Scissors
> A small amount (approximately 1 quart) of waterproof paint; this paint is available in any hardware store for use in waterproofing basement walls or sealing concrete fish ponds.

Next, I offer the step-by-step procedures.

> Constructing the aquarium consists of the following 8 steps:
>
> 1. Removing the inner components of the monitor.
> 2. Cutting an opening in the top.
> 3. Installing the clear plastic "window" onto the front of the monitor.
> 4. Sealing the back of the aquarium with fiberglass.
> 5. Reassembling the monitor.
> 6. Final sealing.
> 7. Coating the aquarium with waterproof paint.
> 8. Testing for water-tightness.

There are some flaws in this writing, beyond prolixity. I tend to use too many parenthetical phrases instead of footnotes or separate sentences. Some of the phrasing is awkward. I really don't know if this manual was effective because no one ever contacted me to indicate a successful execution of the project.

After a year, the original fish tank started leaking irreparably. I hope that the auctioned aquarium fared better. In any case, you can now buy monitor aquariums on the Web, so I am finished building them.

6.2.3 Vignette: Operational Instructions for Krav Maga

Now let us turn to writing training procedures or operational instructions. This example involves describing how to perform certain techniques from the Israeli system of self-defense called Krav Maga, which is one of my favorite pursuits. Here are some questions and my answers (in italics) for an instructor certification exam that I took:[3]

1. List the basic method for breaking down a technique when teaching:
    a. *Identify the immediate threat*
    b. *Discuss how to remove the threat*
    c. *Discuss the simultaneous counterattack*
    d. *Discuss follow-up combatives and variations*
2. In a headlock from the side, explain what part of the technique addresses the immediate danger.

The immediate threat is to the airway/neck, so turning/tucking the chin reduces this threat.

3. Explain the difference (if any) in the techniques for defending against a bar arm-type headlock from behind and a cardioid-type headlock from behind?

The threats are slightly different (airway versus blood choke), but there is essentially no difference in the defense technique. The key is to start either defense as soon as possible and, especially in the rear-naked choke (cardioid), to prevent the attacker from getting the full choke on (which is much harder to defend).

4. Explain why it is important to train from a poor state of readiness. How does this contribute to a better self-defense system? How does this contribute to the creation of techniques themselves?

In Krav Maga we always want to train in situations of maximum disadvantage (fatigued, injured, surprised, etc.). Techniques that don't work in these situations are not nearly as valuable to us, because we need to know that we can apply defense techniques anytime. After all, we may be attacked when we are in such a situation.

And when we are tired, surprised or injured, we are likely to revert to our instincts. Moreover, training in a disadvantageous situation makes us more formidable if we are attacked in an "ordinary" situation of readiness.

Therefore, techniques for Krav Maga were created (and continue to be adapted) to be compatible with these disadvantageous situations. All techniques include simple movements that build on our own instincts, and the system is always being adapted to improve the defenses.

5. How does the technique choke against the wall change if there are multiple attackers or a concealed weapon?

The initial part of the technique doesn't change. You still have to escape the initial threat (the choke) and get away from the wall. But once you get out from the wall, your combatives and strategy are dictated by the situation. For example, if there is a single attacker and he goes for a weapon, you need to neutralize that threat. If an attacker reaches for his pocket, for example, I would trap his hand in the pocket while simultaneously counterattacking with my other hand. If he has already drawn the weapon, I would use the appropriate defense for that situation.

On the other hand, if there are multiple attackers, my strategy is dictated by the location of those attackers and the nearest exit. I need to get between the attackers and the exit, and try to keep only one attacker in front of me (i.e., "stack" them). I may need to use the initial attacker (the choker) as a shield against the other attackers. Or, I may be able to escape the choke, get away from the wall, and get to an exit. If there are multiple attackers and weapons, well, then there are so many variables to deal with. I'd go for the most imminent threat first, try to neutralize him, then use that attacker as a shield as I fight to get to the exit.

6. Create an original training drill to test a technique under stress.

To train a choke against the wall under stress I would form groups of three. One person starts by holding a tombstone pad. An attacker awaits at a wall 20–30 feet away. The defender starts by hitting the tombstone pad as fast as they can for a ten count (the holder counts). Then the defender runs to the wall. The attacker chokes the defender against the wall either with his hands or forearm. The defender needs to defend, counter-attack, escape the wall, and then run back to the pad. Then the defender hits the pad again for a ten count and returns to the wall where the attacker awaits. The defender needs to defend, counter-attack, and escape five times.

To rotate, the defender becomes the pad holder, the attacker defends, and the previous pad holder becomes the attacker.

This kind of procedural writing, which does not involve tools, materials, or a formulation of some kind, can be difficult to write. It is often helpful to go through the steps of the procedure before, during, and after you write the first and subsequent drafts. I suggest having more than two people read and test the procedures after you have completed your own reviews.

6.2.4 Vignette: Recipe for Fennell Pasta

Finally, cooking and baking recipes are kinds of technical procedures. Think of the similarity to chemical formulation or mechanical assembly instructions. In some ways, developing software has similarities to cooking—the focus on quality ingredients (components), processes and tasting (testing) at each stage. I once developed a proposal to give a lecture about the similarities while simultaneously preparing a meal for the audience. At the end, we would eat the meal and continue the discussion. I had even secured funding for this project, alas, other activities took precedence.

The following recipe is one that is modelled after a similar meal I enjoyed at some long-forgotten restaurant. Being raised by a Sicilian mother, I can tell you this is authentic Southern Italian cuisine. I hope you enjoy it.

Penne Pasta with Sausage and Fennel[4]
Serves 6

Ingredients

1 pound penne pasta
1 pound hot or sweet Italian sausage (for a more intense fennel flavor use fennel sausage) cut into discs
1 large fennel bulb cut into small pieces (about the same size as the penne)
1 small onion chopped
1/2 cup sun dried tomatoes cut into small pieces
2 tablespoons olive oil + additional for finishing
¼ cup chopped fennel fronds
¼ cup shredded Pecorino Romano cheese

Preparation

Heat 1 tablespoon olive oil in a sauté pan
Fry sausage in olive oil until cooked
Remove sausage to plate
Drain excess fat from frying pan
Add 1 tablespoon olive oil to pan
Add fennel and onion to pan and cook until tender (about 6 minutes)
Return sausage to pan
Add sun dried tomatoes and cook for about 3 minutes
Add pasta water as needed to sauté pan to keep ingredients moist
Add salt and pepper to taste
Add cooked, drained pasta to sauté pan and toss
Add cheese, toss
Transfer to serving bowl
Add additional (high quality) olive oil to taste (I add as much as ½ cup) and toss again
Garnish with fennel fronds and serve

In authentic Italian cooking, quantities are approximate and you can adjust ingredients to taste and availability. For example, if you don't like fennel, substitute celery. You can use chicken instead of sausage or you can make the recipe vegetarian by using tofu (my mother would cringe at this, however).

If you have favorite recipes I hope you are recording and sharing them. See Appendix B for a recipe template.

6.3 Proposals

A proposal is a document that is submitted to a funding agency, an oversight committee, management, or some other entity that can provide permission, funding, or substantive support for some planned activity. The memo to my dean proposing the computers in film series (see Section 5.5) was an example of an informal proposal.

Formal proposals require much more information. Government agencies have elaborate requirements for their proposals. Space does not allow inclusion of a complete example of this proposal type, but here is a modified proposal template for a U.S. government agency, which also can be adapted for use in situations when the funding agency does not require a specific proposal format:

Title

Problem Statement

Hypotheses/Objective

Scope and Limitations

Definition of Terms

Abbreviations

Background

Uniqueness of the Research

Potential Contribution

Direct Application

Potential for Technology Transfer

Methods and Procedures

Success Criteria

Qualifications of the Research Team

Organizational Center Initiative History

Deliverables/Schedule

Estimated Cost

Other Funding Sources

Many private foundations and corporations have simple proposal formats, while some offer none at all, thus leaving you to organize your proposal as a clear and well-structured document.

6.3.1 Vignette: Grant Proposal

As an extended case study, consider the following grant proposal for the study of software requirements and design that I submitted to a company a few years ago. I have left out some of the fields for brevity, and also I have omitted some of the identifying and budget information:[5]

Title: A Study of Software Requirements and Design

Specification Practices Problem Statement:
There is a wide range of techniques available for software require- ments elicitation, specification, and design. It is currently not known which techniques are used across the many NASA internal and subcon- tractor software development groups, how these techniques are used, and whether they are perceived to be successful.

Hypotheses/Objective:
We plan to research how software requirements and designs are specified across NASA internal and subcontractor groups to uncover best practices and to disseminate this information. The purpose of this research is to improve the practice of software requirements and design specification, to seek to obtain a level of uniformity, and to improve the potential for software reuse and overall software quality.

Scope and Limitations:
The study excludes non-NASA and non-NASA-affiliated entities and also excludes NASA entities specifically excluded at the request of NASA.

Definition of Terms:
Software Requirements Specification: The set of documents contain- ing a complete, consistent, correct and verifiable functional and non- functional description of a software system.
Software Design Specification: The set of documents that completely describe how a software system is to meet the requirements set out in the Software Requirements Specification.
Software Requirements Elicitation: The process and practice of determining the functional and nonfunctional requirements of a soft- ware system for the purposes of developing the Software Requirements Specification.
Best Practices: The collection of rules, procedures and behaviors that are known to lead to efficient production of software include the elicita- tion of requirements, followed by writing and validating the Software Requirements and Design Specifications.

Background:

Based on the experience of the principal investigator and of the project consultants, one of whom has been working with NASA for many years, it is clear that there is a wide range of software requirements and specifications techniques being used across the many NASA software development organizations, and by vendors providing contracted software to NASA. This wide variation is no different from what is found in the software development industry at large.

There are many reasons for the diversity of methodologies used in software specification and design. These include the background and education of the engineers, the application domain, the corporate culture, and the existence of prior documentation which is to be reused. In addition, wide variation in software requirements and design specification methodologies makes software reuse difficult and causes dramatic variation in the software product as well as in the productivity and versatility of software engineers.

Poor software specification and design combine to form the leading cause of software failure, delays in production, and cost overruns. Too often, new techniques are introduced for software requirements and design specification that are intended to mitigate these problems. Unfortunately, because of insufficient training, lack of desire, poor follow-through, and/ or pressures to complete the project quickly, these techniques are often adopted incorrectly or half-heartedly. Yet the impression persists that the problems of poor specification and design have been solved.

Uniqueness of the Research:

There are virtually no published results of surveys of Software Requirements Specifications or of Design Specification practices. A recent search on ten of the most applicable abstract indexing services yielded only two related published works. The first was a survey of requirements methodologies used in the design of databases [Batra] and the other a survey of techniques for the design of software platforms [Johannson]. Further, we know of no survey data specifically covering NASA software project teams or for NASA external software vendors.

The consultants for this project have already conducted research on Software Requirements Specification practices for a broad base of more than one hundred companies in the Delaware Valley. The preliminary results indicate that there is a disconnect between mandated requirements specification practices, implementation of those mandates, and perceived benefits. We believe that this may also be the case for NASA software development groups.

We intend to adapt and apply the prior survey to assess the processes and perceptions of NASA internal and contracted software groups. We will further extend the study to include Software Design Specification practices. The data already collected for the Delaware Valley Companies will act as a "control group" with which to compare the results for NASA internal and external software development groups. We will also compare and contrast the practices for NASA internal groups versus NASA vendor software groups.

Potential Contribution:
This research will have a major impact on software reusability and consistency of documentation across the broad range of NASA centers and vendors and of flight software and support software. The research should lead to improved best practices for all NASA software organizations. Ultimately, it should lead to increased software reliability, maintainability, productivity, and cost savings.

Direct Application:
The proposed research will support all ongoing software development efforts and provide reference data and best practices for software project managers and practitioners. A pilot study has already been conducted by the project consultants for a group of more than 100 companies in the Delaware Valley.

Potential for Technology Transfer:
There is tremendous potential for technology transfer. Since there is little data available on prevailing practices in developing Software Requirements and Design Specifications, these results will be of great importance to software practitioners in all applications domains. There is a great likelihood that the results of this project would be applicable to a whole range of current NASA projects and contractor projects.

Methods and Procedures:
This research depends on intensive upfront work and follow up in succeeding years. This work will involve visits to NASA and contractor sites and intensive telephone and email contacts. In the first year, we will construct the Web based survey instrument. To do this we will assess existing survey instruments including the prior instrument used by the consultants for the Delaware Valley "control group." We will interview various NASA internal and external software development groups. We will construct an email list of survey participants. We will then invite the pool to participate in the survey and we will send various reminders to maximize participation. We will collect data and encourage participation for several months. Upon termination of the data collection phase we will commence data analysis. Finally, we will prepare a report and set of recommendations. Our recommendations will include next steps for the follow up research to be conducted in year two. We will present our findings to NASA and in various proceedings and journals as appropriate.

In the second year, we will focus our investigation on the actual participants from the year one survey (those who provided survey data). We will study selected Software Requirements and Design Specifications documents, manuals and procedures. We will conduct focus groups and interviews (either in person, via telephone or via email). Our objective is to further understand and explain the results of the first year. We will match reported with actual practices and identify any disparities. We also intend to further quantify best practices and to intensively disseminate those findings throughout NASA. This will involve travel to NASA and contractor sites. We will also construct the final research focus for the third year. We will present our findings to NASA and in various conferences and journals as appropriate.

In the third year, we will conduct additional focus groups and follow up studies. The purpose is to see if the findings and best practices disseminated in years one and two have begun to penetrate the behaviors of the software practice groups. The study methodology will be similar to that of year two. We will continue to work with those software groups that find our results beneficial. We will continue to disseminate our finding to NASA software groups. We will present our findings to NASA and in various proceedings and journals as appropriate, including recommendations for future work and initiatives.

Success Criteria:
Success will be measured by collection of statistically significant survey results and meaningful focus group results and by identification of best practices and areas for improvement. Furthermore, we expect to develop improved standards for Software Requirements and Design Practices, and signs of adoption of the best practices in those software practice groups that are identified as most needing improvement.

References:
[Batra] Dinish Batra, "Consulting Support During Conceptual Database Design in the Presence of Redundancy in Requirements Specification: An Empirical Study," *International Journal of Human-Computer Studies,* Vol. 54, 2001.
[Johansson] Enrico Johansson et al., "The Importance of Quality Requirements in Software Platform Development — A Survey," *Proceedings of 34th Hawaii International Conference on System Sciences,* 2001.

Unfortunately, this proposal was not funded—I think for reasons beyond the quality of the proposal. There is tremendous competition for grant monies, and political influence is often more important than persuasive writing. Government funding agencies tend to require a very specific format for proposals and will provide a template. Private foundations often have less restrictions on the proposal format. Appendix B has a generic template for a research grant proposal, which can be helpful in writing a proposal that can then be fine-tuned to a funding agency's format.

6.3.2 Vignette: Proposal for Consulting Services

Sometimes a proposal to a company can be much more informal, as in consulting services. Here is a disguised sample proposal that I submitted to a company to deliver a course.

Proposal for Consulting Services

Agreement made this day February 6, 2017 between

MaggieTex Systems, the Client, and **Dr. Phillip Laplante**, the Consultant, for consulting work by Dr. Phillip Laplante, the Course Leader.

1. **Preamble.** The responsibilities of both the Consultant and Client are outlined in this Agreement. Each has obligations to one another, which when fulfilled in an atmosphere of mutual respect and cooperation, will yield benefits to all concerned.

2. **Course Delivery.** Real-Time Systems Design and Analysis, the Course, is a three-day lecture-style course. The Course will be delivered at the Client's site in Cambridge, Massachusetts on three consecutive days from March 17, 2017 to March 19, 2017.

3. **Client Obligations.** Client shall provide the following items for use by the Consultant in the delivery of the course:

   A lecture or meeting room capable of seating all course attendees

   An overhead display projector and screen

   The Client will also pay the Consultant under the terms of section 5, "Payment Terms."

4. **Consultant Obligations.** The Consultant will deliver the Course as described in section 2, Course Delivery, along with one copy of the course notes. The Client is authorized to photocopy the course notes, one set per attendee. The Consultant will ask attendees to participate in a course evaluation. The results of that evaluation will be shared with the Client. However, payment under section 5 is not contingent upon the results of the evaluation. Consultant will retain copies of all surveys and may use comments contained therein for promotional purposes.

5. **Payment Terms.** The Consultant shall be paid 400 shekels of gold for the delivery of the course. Payment will be due 30 working days after completion of the course. Late payments will incur an interest charge of 1.5% per month. Client is only authorized to copy Course notes, one copy each, for attendees of Course. Additional copies may be made at a cost of 1 shekel per set, or purchased from the author at 2 shekels per set.

6. **Course Changes or Cancellation.** Neither the Consultant nor the Client may reschedule or cancel the Course without the agreement of the other party except that:

   In the event of serious illness on the part of the Course Leader, notice will be given to the Client and the course rescheduled at mutual convenience.

   Client may completely cancel the Course up to 30 days prior to the start of the Course delivery without consequence upon simple notification of the Consultant.

   Client may cancel the course 10 days prior the start of the Course delivery, but will be charged 40 shekels.

7. **Interpretation.** The services described this Agreement constitute the entire agreement between the parties hereto, and supersede all prior verbal or written discussions and agreements.
This Agreement shall be construed in accordance with the laws
of the Commonwealth of Pennsylvania and shall be deemed to
have been accepted in said state. It may not be changed orally.
Any controversy arising out of or relating to this Agreement or
the breach thereof shall be settled by arbitration in Pennsylvania
in accordance with the rules of the American Arbitration
Association, and the award rendered by the arbitrator may be
entered in any court having jurisdiction thereof.

All proposals are legal documents—they constitute a commitment on the
part of both parties, but this informal proposal really looks like a contract,
and it is one. You should consult with an attorney before submitting any
contract to a client.

Here is the cover letter to go with the proposal for technical services:

Dear Dr. Lee:

This is in reference to the Real-Time Systems Design and Analysis
course that I will be delivering at your site on March 17–19, 2017.

First, thanks for sending the background information on MaggieTex
Systems. I am sure that the course is very much aligned with your applications areas.

Please find enclosed one copy of the course notes to accompany my
book, *Real-Time Systems Design and Analysis, Fourth Edition,* which I understand you are purchasing for course attendees, and a course evaluation
form. You are hereby authorized to make one copy of the course notes
and the course evaluation form for each person attending the course on
March 17–19. Additional copies of the course notes may be made for non-
attending personnel provided you notify me in advance of the number
of copies to be made. I will then invoice MaggieTex Systems 1 shekel of
gold per copy for the additional copies.

I am looking forward to our course and to meeting with you and the
other people at MaggieTex Systems.

Please don't hesitate to contact me in the meantime if you have any
questions.

Sincerely,
Phil Laplante

In submitting my proposal, I also included a summary of the course, which
is included here for completeness.

REAL-TIME SYSTEMS DESIGN AND ANALYSIS (3½ hours)

Real-time and embedded systems, which are closely related, are so ubiqui-
tous that they are impossible to avoid. The term real-time has even entered

non-technical jargon. But real-time systems are special and require special considerations to design. Based on the second edition of the bestselling text, *Real-Time Systems Design and Analysis: An Engineer's Handbook,* this course provides an introduction to real-time systems and the real-time problem.

> BENEFITS/LEARNING OBJECTIVES
>
> > This course will enable you to:
> >
> > > Identify the unique characteristics of real-time systems
> > >
> > > Explain the general structure of a real-time system
> > >
> > > Define the unique design problems and challenges of real-time systems
> > >
> > > Apply real-time systems design techniques to various software programs
>
> INTENDED AUDIENCE
>
> > This course is ideal for newer software engineers or experienced software engineers who have never worked in real-time or embedded software environments. Managers of projects involving real-time systems will also benefit.

All proposals involve commitment; therefore, you must be extremely cautious in what you put in writing. There may be a temptation during the proposal phase to exaggerate your abilities or overreach your goals. You must curtail these urges lest you find yourself winning a proposal on which you can't deliver. Appendix B contains a template for a consulting services proposal.

6.4 Panel Sessions

You may be asked to lead a technical panel discussion, forum, or roundtable. In many ways, leading a discussion panel is like leading a meeting. You need to control the pace of the meeting, keep participants on track, and stimulate discussion when needed. In preparing for these kinds of events, you must write a planning document, which is essentially a list of questions or discussion items.

I am frequently asked to lead or participate in discussion panels at various conferences. Whether I am the leader or a member of the panel, I always prepare notes. Here is a set of planning notes and questions for a panel discussion on computer security that I was asked to lead:

QUESTIONS FOR ITEC COMPUTER SECURITY PANEL
October 31, 2018

1. Philadelphia, Pennsylvania, Delaware, New Jersey and the federal government are rapidly increasing the number of services that they provide electronically. How do you go about weighing which services to put on line and which not to provide because of security concerns?
2. What are the most serious threats to the security of your data and communications systems?
3. Where are the major cyber threats coming from?
4. What precautions are you taking to protect your systems?
5. Can you tell us about any recent incidents in which a threat had been identified and neutralized?
6. Has the 9-11 terrorist attack changed the way that you think about data and communications systems security?
7. When building secure systems, there are always privacy issues, for example, issues surrounding the use of private data in the process of authenticating and recording transactions. How do you balance security against privacy?
8. Some organizations employ professional hackers to attempt to identify system vulnerabilities. What do you think about this practice?
9. In general, how do you recommend companies improve the security of their systems?
10. No matter what type of electronic security systems that are put in place, the "man in the loop" often represents the soft spot. How do you deal with this concern in your organizations?
11. How would you go about evaluating the vulnerability of third-party software before adding it to your overall systems architecture?
12. What are you looking for in vendors' software to assure you that their product is not going introduce a weak link in the overall systems security?

I often use my notes from a panel session to write some kind of paper that I can distribute as minutes or convert into some published work.

6.5 Strategic Plans and Planning

While it can be quite boring, strategic planning is an important activity for any organization. Often taking weeks or even months, strategic planning involves assessing positive and negative forces with respect to the internal status of the organization and the environment. Strategic planning should involve everyone in the organization, but it usually does not. I have participated in many strategic planning sessions, and have led them for nonprofit and for-profit organizations.

The primary artifact of the strategic planning activity is the strategic plan. The strategic planning process isn't a one-time activity—the strategic plan is used as a mechanism for execution, a compass to help make strategic decisions, and as an "annual report" for marketing purposes, recruiting staff, or fundraising.

The template for a strategic plan can be found in Appendix B. Let's look at the components of a strategic plan in some detail.

6.5.1 Executive Summary

The executive summary provides a short synopsis of the overall contents of the strategic plan. The summary allows the reader to quickly identify the key themes and recommendations of the report. Here is a simple mission statement for a fictitious university:

Executive Summary
This plan sets the framework, tone, and direction for Piedmont University for the next three years. It proceeds from the previous strategic plan (2012–2017) and charts a course to preserve academic excellence, innovation, and distinctiveness within our market of Southeastern Pennsylvania. This plan was developed by a committee, which included representation from across the university. The committee sought input through group and individual meetings with all members of the community. This process created a plan that reaffirms our vision of technical education and our mission as a technical university while setting out distinct plans to place Piedmont University in the best position to compete in the current economy and educational environment.

Today, Piedmont University has two campuses offering undergraduate degree programs in education, technology, and management, with approximately 4400 students enrolled. In addition, Piedmont University provides certificate programs and other professional development opportunities to 1200 students.

Specifically designed for busy adults, programs are offered in day, evening, and weekend classes. Conferencing services and facilities are also available. After a number of growth years, we are now faced with the challenge of a downturn in enrollment. Planning for the future requires Piedmont University to carefully examine its strengths, weaknesses, processes, and competitors and to refocus its members on core values and business processes. We believe the result is a plan that builds on existing strengths, calls for change where needed, and establishes a framework for innovation, cooperation, and achievement.

In particular, the plan calls for:

- A broad-based one million-dollar marketing program
- The addition of two new programs, one in information systems and one in healthcare informatics

- Hiring 15 new faculty members
- Consolidation of the telecommunications and computing departments

These steps will lead to an overall increase of 300 students a year and a net revenue increase of five million dollars per year, at a projected expense of five million dollars per year.

The executive summary is written to excite the readers and prepare them for the details that follow in the remainder of the plan. The executive summary is always written last; otherwise, you will need to constantly revise the summary as the contents of the strategic plan change during its writing.

6.5.2 The Mission Statement

The mission statement is a description of purpose for an organization, and allows stakeholders to weigh the importance of various initiatives by asking the question, "How does that activity serve the mission?" A good mission statement is compact and powerful. One of the best mission statements is associated with the Starship Enterprise from the original *Star Trek* series. That Mission Statement was [Asherman 1987]:

To explore strange new worlds, to seek out new life and new civilizations, to boldly go where no man has gone before.

This statement is clear, compelling, and inspiring. Beyond those qualities, it is useful. I recall several episodes where the crew's behavior was guided by the mission statement.

I don't advise that you revisit the mission statement during strategic planning because you can get tangled in turf wars and minutiae. Developing the mission statement should be a separate activity and should have been completed before beginning the strategic planning process.

Let me offer another example of a mission statement for a fictitious, small technological school named "Piedmont University." Here is its Mission Statement:

Empowering people through technology.

You may not like this mission statement but it is a good one. It is very brief and it is certainly one that will not need to be changed frequently to meet the demands of changing technology or societal norms. It can be used as a guidepost for decision making. For example, if someone were to attempt to introduce a new degree program in literature, it would likely be rejected because it does not fit the mission statement.

6.5.3 SWOT Analysis

A Strengths, Weaknesses, Opportunities, and Threats (SWOT) analysis is a study of internal and external factors and positive and negative forces. The SWOT analysis framework can be illustrated by a two-dimensional array (Table 6.1). Opportunities are based on positive external forces and threats upon negative ones. Strengths are based on positive internal forces and weaknesses on negative ones.

To conduct a SWOT analysis, a comprehensive environmental scan and internal self-studies are conducted. Then the results of these studies are organized along the dimensions of Table 6.1.

Here is an example of a partial SWOT summary for Piedmont University:

Strengths:

    Small class sizes

    Excellent faculty

    Beautiful campus

    Ample investment portfolio

Weaknesses:

    Low enrollments

    Aging infrastructure

    Dependence on adjunct faculty

    Low graduation rate

Opportunities:

    Growing market for healthcare informatics degree

    Growing market for information systems degree

    Partnership opportunities with two-year colleges

Threats:

    Emerging competition

    Economic downturn

    Anticipated key faculty retirements

TABLE 6.1

SWOT Analysis Framework

	Positive Force	Negative Force
External Factor	Opportunities	Threats
Internal Factor	Strengths	Weaknesses

Of course, each of the bullet points summarizing these strengths, weaknesses, opportunities, and threats are elaborated and expanded in the narrative of the SWOT section of the strategic plan.

6.5.4 Competitive Market Analysis

The market analysis examines the market demographics, customer characteristics, economic trends, and competition. Any book on writing a business plan or a good marketing textbook will provide thorough guidance on conducting a market analysis.

6.5.5 Goals, Objectives, and Strategies

The mission statement is a very high-level statement of what an organization stands for. But more details must be provided to articulate that mission. The mission statement is articulated through strategic goals that describe specific desired outcomes. Goals are linked to strategic objectives, which can be measured. Thus, goals are operationalized through strategies. The hierarchy is shown in Figure 6.1.

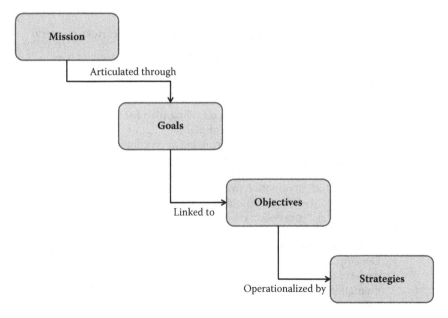

FIGURE 6.1
Hierarchy of concepts for strategic planning.

Goals are broadly scoped and enduring outcomes. Here are some sample goals for Piedmont University:

Goal 1:   To create an inclusive environment for teaching and learning

Goal 2:   To foster an environment for financial sustainability

Goal 3:   To be a positive force in the local community

Good strategic objectives are detailed through strategies. Strategies need to be:

Mission consistent

Actionable

Measurable

Persistent

Embraceable

For Piedmont University, here is Strategic Objective number 2.3 under Goal number 2: "To foster an environment for financial sustainability":

Objective 2.3: Establish mechanisms for revenue enhancement and cost control.

Objective 2.3 has nine strategies (and corresponding metrics), namely:

A. Develop a process for forecasting tuition revenue based on historical data and well-defined recruitment-to-classroom pipeline model (metric: date of publication)

B. Institute a budgeting process that directly ties funding requests to the strategic plan (including recycling plan for high-priority objectives) (metric: publication of plan)

C. Institute a process for lead sharing between continuing education and resident instruction (metric: date of publication of plan)

D. Increase average class size (metric: average class size)

E. Increase student retention (metric: student retention rates)

F. Increase conversion of qualified applicants to registered students (metric: conversion ratio)

G. Enhance non-tuition revenues (e.g., conferences and institutes, grant dollars) (metric: monies)

H. Develop and implement a continuing education plan, which projects revenue increases (metrics: date of publication, implementation details)

I. Establish a forward-looking faculty senate finance committee to assist in understanding and communicating major financial issues (metric: date of establishment)

Strategic planning involves numerous workshops, studies, interviews, and data-gathering activities. Therefore, a wide variety of reports accompany the strategic planning activity.

6.5.6 Budget

A convincing budget is just as important as the technical and organizational content of the strategic plan. The budget can include both best-case and worst-case scenarios, as well as budget management plans, contingency planning, etc. A detailed discussion of the preparation of budget documents is beyond the scope of this text.

6.6 Problem Reports

Problem, incident, or trouble reports are a means of communicating issues arising in the operation of a system to its maintainers. Software defect (bug) reports and defective product return reports are examples of problem reports. More extensive incident reporting is required when major systems fail.

Problem reports contain the following elements:

Name of the problem

Product identification information

Description of the suspected problem, including as much detail as possible

Date the problem was discovered

Name of person who discovered the problem

Condition under which the suspected problem occurred

Description of expected results

Actual results

A rating of the problem severity

Contact information

The problem report should be comprehensive and clearly written—this will help the maintainers quickly duplicate and fix the problem. A sloppily written report is both unhelpful and liable to repel the maintainers.

The manner in which you convey the problem information in the problem report is crucial in helping maintainers decide if the problem is already known or if it is a new one. This reduces bug-report duplication and assists in repair assignment. Duplicate error reports, for example, can raise the maintainer's sense of urgency, even if the error is trivial. Users can also identify an error incorrectly as a duplicate of another already reported, and they can employ inconsistent terminology, thus preventing maintainers from recognizing duplications [Laplante and Ahmad 2009]. My experience suggests that the more time you take to carefully prepare the problem report, the more likely you are to discover that the problem was not really due to system malfunction, but was linked to your own misunderstanding or failure to follow procedures.

Here is a quick example problem report of a suspected problem with my electric toothbrush system, which incorporates an ultraviolet light sanitizing feature.

Name of the problem: The ultraviolet sanitizer does not work.

Product and Model Number: Fred's Toothbrush System, Model # 3455

Description: Sometimes when I press the button to turn on the sanitizer, the indicator light does not go on and the expected ultraviolet glow does not appear through the observation window. The problem only occurs intermittently. I purchased and replaced a new light bulb, but the problem persisted.

Date: 10/12/2017

Contact: Phil Laplante at plaplante@psu.edu

Tools such as the open-source Bugzilla defect-tracking system can be quite helpful in collecting and analyzing error reports. Tracking systems act as a database for reported problems. Problems are denoted as "new" when reported, "assigned" when assigned to a maintainer, and "resolved" when the report is closed. Tracking systems greatly assist maintainers in problem triage and effort allocation.

6.7 User Manuals

There is very little published work on best practices for writing user manuals. Most information focuses on generic technical writing or on system usability. But it is quite important to write high quality users manuals both to preserve the reputation of the product and for safety. As noted in Chapter 1, poorly written user manuals can lead to disaster [Wong et al. 2017] (see Vignette 6.7.1).

Describing a well written user manual is difficult—it is easier to identify a badly written one. So let's start by discussing the characteristics of badly written user manuals. They're poorly written and difficult to follow. Necessary illustrations, an index, and key information are missing. The instructions force you to flip from page to page. There is no discussion of what to do when something goes wrong. Sometimes, there are funny mistranslations, malapropisms, and incorrectly used heterographs.

A well-written user manual exhibits the opposite of these bad characteristics. That is, the manual is well written and easy to follow. It is well illustrated and all the information you need can be found easily, either through a table of contents or the index. There is a comprehensive list of common problems and what to do about them and contact information in case you need help or have questions.

User manuals should be tested as part of final system acceptance testing. Remember the 5 Cs of good technical writing from Chapter 2? That is, good technical writing is correct, clear, complete, consistent and changeable. These qualities can be used as guides while testing user documentation. For example, to test correctness, a tester should follow every procedure in the manual and verify that they are correct. Clarity could be shown through measures of the language used (e.g., Flesch-Kincaid reading ease indicator less than a certain value, say, 8), inclusion of effective graphics and navigational aids for the reader (see Chapter 7), For completeness, it should be proven that ever feature in the system is fully documented. Consistency would pertain to use of language and writing style as well as depth of discussion (for example, each feature should have the same level of detail in its discussion). The changeable quality could be met through online hosting of errata and updates to the manual that can be downloaded.

Finally, here are some guidelines for producing good user manuals. Ideally, they should:

Involve the technical writer in all phases of product development so that they can write about the why a certain feature exists as well as to how to use the feature.

Write the user manual at the end, not during the process of product development as features will change until the final system delivery.

Keep the language very simple and straightforward.

Not use jargon, but you should include a glossary where any standard language is used in a nonstandard way.

Have the manual copy edited by another technical writer.

Have one or more persons, who did not write it, test the user manual.

Provide frequently asked questions and troubleshooting sections.

Provide a quick start guide if applicable.

Always include a table of contents and index.

Provide a hosted site with errata and updates to the manual that can be downloaded.

More research needs to be conducted on creating and measuring good user documentation patterns and antipatterns. If you are involved in writing a user manual, a template for a generic one is provided in Appendix B.

6.7.1 Vignette: Disaster from a User Manual

In a notorious incident a radiation therapy device, the Therac-25, accidentally gave massive overdoses to cancer patients. In six reported accidents between 1985 and 1987, patients were given nearly 100 times the specified amount of radiation, apparently causing three to die. Investigations determined that the accidents were caused when a certain sequence of keystrokes typed on the system terminal created an overdose situation and the failsafe mechanisms malfunctioned. While several factors contributed to the catastrophe including, insufficient integration testing, poor systems documentation, and sloppy development practices, ambiguous user manuals were also blamed. In particular, the Therac-25 documentation did not adequately explain certain error codes that could be generated and on an instruction sheet for the machine "dose input 2" could ambiguously mean that either too high or too low a dose had been delivered [Leveson and Turner 1993]. Clearly the Therac-25 documentation failed on at least two of the 5 Cs (correctness and completeness) and probably didn't meet the other desired qualities of a good user manual.

6.8 Exercises

6.1 Write a step-by-step procedure for making your favorite recipe, including all ingredients and utensils needed.

6.2 Write a step-by-step procedure for doing a typical load of laundry from washing through folding and hanging.

6.3 Write a proposal to your current boss (or a hypothetical one) to establish a new department of "leisure" to provide entertainment activities for employees during break and lunch times.

6.4 Write a problem report for some unfortunate happenstance in which you were involved (e.g., a flat tire or missed appointment).

6.5 Evaluate this mission statement for a hypothetical automobile manufacturer: "We make great cars."

6.6 For the mission statement in Exercise 6.7, state one goal, three objectives that address that goal, and for each goal, two strategies.

6.7 Select a user manual for a product that you use regularly. Conduct a 5 Cs analysis of the user manual.

6.8 Find a humorous error or errors in some user manual. Rewrite that portion to remove the levity.

Endnotes

1. Excerpts reprinted with permission from Phillip A. Laplante, *Easy PC Maintenance and Repair*, second edition, Windcrest/McGraw-Hill, Blue Ridge Summit, PA, 1995. © 1995 McGraw-Hill.

2. This excerpt represents another interesting case in permissions. Although I am reproducing a substantial amount of text from an article I published in *Aquarium Fish* magazine, I did not have to explicitly request permission because "All rights revert back to you [the author] after we publish your article/photos" (http://www.fishchannel.com/writers-guidelines.aspx, accessed December 2, 2010). Since the publication of the first edition of this book, the magazine had gone out of business.

3. Thanks to my instructor and friend Ernie Kirk of Premier Martial Arts for allowing me to reprint these.

4. I have no idea what the nutritional information is for this dish, eat at your own risk.

5. Permissions note: Because I wrote this proposal, and it was not funded, as the writer I own the copyright to this material. If the proposal had been funded, then the copyright would have reverted to the funding agency or company.

References

Asherman, A., *The Star Trek Compendium*, Titan Books, London, 1987.

Bucholz, S., *The Unofficial Harry Potter Cookbook: From Cauldron Cakes to Knickerbocker Glory—More than 150 Magical Recipes for Muggles and Wizards*, Adams Media, Avon, MA, 2010.

Laplante, P. A., *Easy PC Maintenance and Repair (Second Edition)*, Windcrest/McGraw-Hill, Blue Ridge Summit, PA, 1995.

Laplante, P. A., Computer monitor to aquarium, *Aquarium Fish,* 51–59, October 2003.

Laplante, P. A. and Ahmad, N., Pavlov's bugs: Matching repair policies with rewards, *IT Professional*, 11(4), 45–51, 2009.

Leveson, N. G. and Turner, C. S., An investigation of the Therac-25 accidents. *Computer*, 26(7), 18–41, 1993.

Wong, W. E., Li, X. and Laplante, Phillip, Be more familiar with our enemies and pave the way forward: A review of the roles bugs played in software failures, *Journal of Systems and Software*, 133, 68–94, 2017.

7

Using Graphical Elements

7.1 Breaking up the Monotony

Dull writing can diminish a reader's attention and understanding. In Chapter 2, I discussed how humor and allegory can be used to enliven your writing and hold a reader's attention. But humor and allegory should only be used in limited circumstances, and even then, sparingly.

A better way to hold a reader's attention is with appropriate use of graphical elements. By graphical elements, I mean anything that is not just text. Examples include figures, equations, photographs, and tables. I like using graphical elements liberally because they can enliven the presentation, enrich ideas, and alleviate the boredom of a dry, text-only exposition. Even bullet lists help to change the reader's pace and by breaking up a long run of monolithic prose. This chapter addresses appropriate uses of these graphical elements in technical writing.

7.2 Modeling Ideas with Graphics

Words, tables, figures, equations, and so on can be used to model any idea. This doesn't mean that modeling ideas with words or pictures is easy—you know that writing can be hard. Creating good charts and writing equations can sometimes be harder. The challenge is to find the right techniques for the ideas that you want to explain.

The front matter of my church's missal lists an "art director," yet there is almost no visual art in the book except three infrequently appearing icons. So, why does the missal need an art director? The answer is that seeking the right mix of text and non-text elements is a work of design. In the missal there are hymns, prayers, music, scriptures, and notes. Finding the best layout for all these elements is a form of art. There are also many fonts used, helping to guide the reader and decorating the book in a subtle way.

If there are too many graphical elements in a piece of writing, the reader becomes distracted as the flow of the narrative is lost. If there are too few graphical elements, the reader's attention can also be lost in overly long textual passages. Many concepts can be illustrated via drawings, mathematics, graphs, lists, and so forth. But how do you obtain the right mix of text and non-text elements in technical writing? And how are these items best arranged?

7.2.1 A Picture Is Worth 1437.4 Words

An old adage is that "a picture is worth a thousand words," but why not 1473.4? Can the value of the information in a picture be measured? A group of researchers tried to measure the impact of graphical elements in scholarly research. Using a modified version of the page ranking algorithm used by Google, they found that papers attracted approximately two more citations for every diagram included. Conversely, equations and photographs seemed reduce the number of citations. While these findings should not be taken literally—the value of any picture depends on its context and really can't be measured—it is certainly true, as the authors of the study conclude, that "illustrating an original idea may be more influential than quantitative experimental results" [Lee et al. 2017].

As an example, consider the following discussion of traffic fatalities for different countries:

> In 1998 there were more than 1,170,000 reported traffic fatalities worldwide. The number of traffic fatalities in poor nations was 8.7% higher than in wealthy nations. China and Africa had the misfortune of leading the way with more than 178,000 and 170,000 deaths, respectively. Of the wealthy nations, those in Eastern Europe fared the best, with only 923 deaths.

This narrative only gives a tidbit of information about worldwide traffic fatalities. To give a complete picture, I would have to go on with numbers in textual form for many paragraphs. In addition to being hard to follow, making the numbers interesting would be challenging without taking great liberties of style that seem unwarranted.

Now consider Figure 7.1. This figure conveys more information in a more compelling way, which is more easily digested than the preceding narrative description. The number of the fatalities per region, as depicted by the boxes, grabs your attention and makes comparisons very easy.

But be aware that graphical representation can be used as a weapon to shock the reader or misrepresent information to advance a hidden agenda. In the case of Figure 7.1, the whole story is not obvious because the chart gives only gross fatalities, not fatalities as a percentage of population. India's population in 1998 was estimated at 984,003,683, while China's population was 1,236,914,658 during the same period (*1998 CIA World Factbook*). Computing the number of traffic fatalities per 1000 people in China and India, we get 0.01% versus 0.02%. India had twice the traffic fatality rate of China in that year. Of course, there are deep underlying factors for this difference—differences in

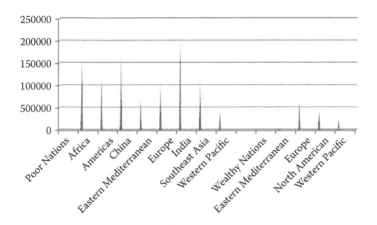

FIGURE 7.1
Worldwide Road Traffic Fatalities — 1998. (Modified from http://www.safecarguide.com/exp
/statistics/statistics.htm)

mass transportation systems, roadways, population density, and so forth—
that can be more easily described in words. Sometimes words are better than
pictures, although usually both together make the most powerful exposition.

7.2.2 Modeling Behavior

When attempting to describe complex behavior, it is usually better to use an
appropriate graphical technique. Unlike natural languages (e.g., English), a
well-drawn diagram is unambiguous.

For example, my home answering machine can be accessed remotely by
dialing my phone number, waiting for the greeting to play, and then entering
a passcode. Once the passcode is validated, the machine tells me that I am
in the main menu. Then I am presented with a number of actions that can
be taken next. For me to describe these subsequent menu options and error
handling would take me a few paragraphs—and I am likely to make a mistake. However, look at the diagram shown in Figure 7.2, which is known as a
finite state machine diagram. It is likely that you can quickly understand the
workings of my answering machine by inspecting Figure 7.2.

Why is it likely that you understand Figure 7.2 more clearly? Because you
are already familiar with the workings of telephone answering machines.

A graphical description may not always serve as an appropriate substitute
for text, particularly when the reader has little or no prior domain knowledge. Here is an example of the problem. One of my hobbies is the martial
art Brazilian Jiu Jitsu (BJJ). I'm not particularly good at it, but I try hard and
mix it with the other systems that I practice (including Krav Maga, which
I already discussed). BJJ has lots of moves and counters to those moves. To
help me train, I have developed a small subset of these moves, which I have
summarized in graphical form in Figure 7.3.

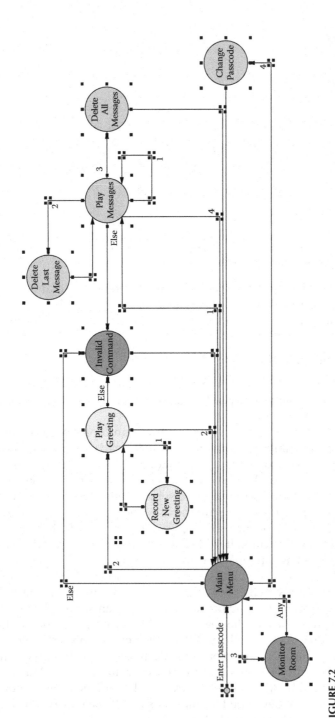

FIGURE 7.2
A finite state machine model for my answering machine.

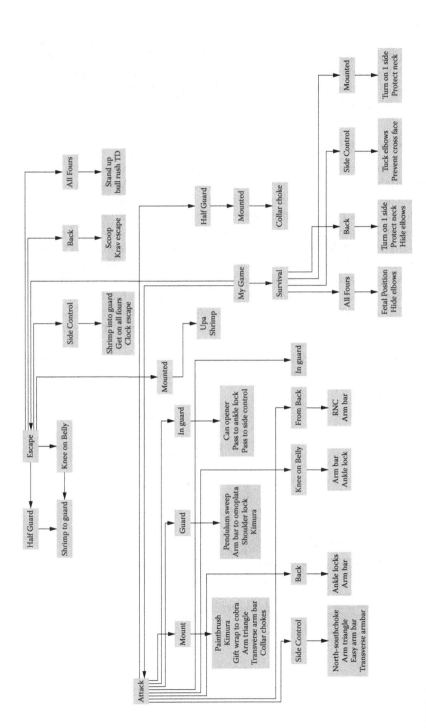

FIGURE 7.3
A summary of certain Brazilian Jiu Jitsu positions and my favorite moves and countermoves.

I am not going to describe Figure 7.3 in words—it would take many pages to do so. But for someone who is familiar with BJJ, this diagram should be understandable. The key to understanding Figure 7.3 is that the underlying specialized domain knowledge between the writer (or artist) and reader are similar. Therefore, when using graphical elements for complex or arcane topics, the necessary domain knowledge must be stipulated or provided.

7.2.3 The Evolution of an Idea

Graphics are a powerful means to show the evolution of an idea from conception to completion. Here is a nice example of what I mean. A few years ago, my wife and I decided to give our house a facelift. We hired an architect who interviewed us about our needs and desires and then came up with several different concept sketches for us to consider. We selected the concept shown in Figure 7.4.

After further consultations and additional research, the architect refined the concept sketch to the drawing shown in Figure 7.5. The architectural drawing was accompanied by floor plans and various other specifications.

After evaluating the costs and making some further adjustments, the architect developed a set of construction drawings for the builder's use. One of these drawings shows the construction details for the footing column support system for the porch, as shown in Figure 7.6.

Because of the level of detail, Figure 7.6 is only interesting to an engineer or construction professional. There were many more construction drawings

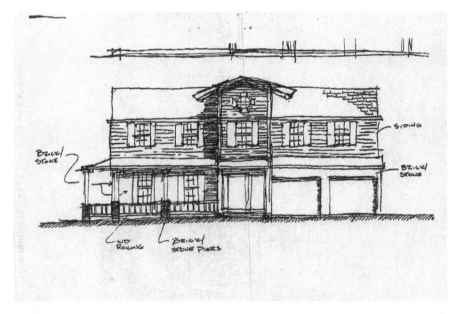

FIGURE 7.4
Architect's concept for a proposed remodeling of my home.

FIGURE 7.5
Architectural drawing (one version) of the proposed remodeling of my home.

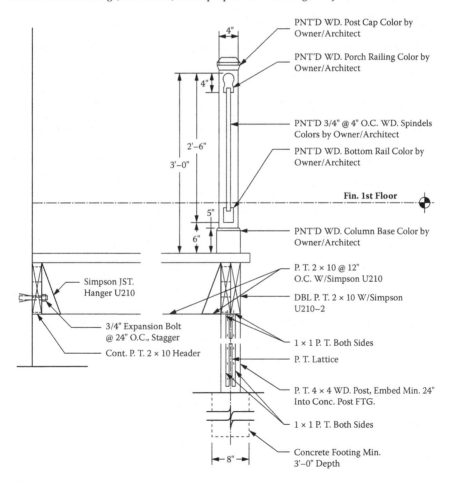

PNT'D WD. Post Cap Color by Owner/Architect

PNT'D WD. Porch Railing Color by Owner/Architect

PNT'D 3/4" @ 4" O.C. WD. Spindels Colors by Owner/Architect

PNT'D WD. Bottom Rail Color by Owner/Architect

Fin. 1st Floor

PNT'D WD. Column Base Color by Owner/Architect

P. T. 2 × 10 @ 12" O.C. W/Simpson U210

DBL P. T. 2 × 10 W/Simpson U210–2

1 × 1 P. T. Both Sides

P. T. Lattice

P. T. 4 × 4 WD. Post, Embed Min. 24" Into Conc. Post FTG.

1 × 1 P. T. Both Sides

Concrete Footing Min. 3'–0" Depth

Simpson JST. Hanger U210

3/4" Expansion Bolt @ 24" O.C., Stagger

Cont. P. T. 2 × 10 Header

4"

4"

2'–6"

3'–0"

5"

6"

8"

FIGURE 7.6
Detailed construction drawing of one small part of the proposed remodeling of my home—the post and support structure for the front porch.

FIGURE 7.7
Photograph of the finished remodel.

for various aspects of the remodel. We gave these specifications to several builders and obtained quotes for the project.

The renovation proceeded, and my wife and I were delighted with the result, shown in Figure 7.7. As a bonus, our renovation project won an award from a regional builders association.

The telling of the remodeling story would have been very difficult—and less compelling—without the use of these figures. But there is a more important point to make. All of these images were a model of the same thing—my home. Each model offered a different level of detail and represented a different point in the conceptual evolution, climaxing with the image of the remodeled home.

7.3 Selecting the Best Model for a Schedule

As in the progression of the models of my home, different models of a project schedule can show increasing levels of refinement. Take a look at the Gantt chart in Figure 7.8.

Gantt charts are a great way to develop and communicate project schedules, which are essential for team-oriented projects (see Chapter 10).

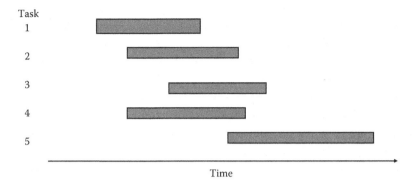

FIGURE 7.8
A simple schedule graphic.

The Gantt chart in Figure 7.8 contains a great deal more information than you might think. The chart depicts subtasks in some project (imagine it has to do with the remodeling of my home) and the relative duration of each task. We don't know the units of time along the time axis, but you can see that Task 5 takes about 50% more time than Task 3. Task 5 is preceded by Task 1 and is concurrent, at least for a while, with Tasks 2, 3, and 4. Some forms of Gantt charts can show dependencies, that is, where one task may not begin until another has been completed.

Figure 7.9 is similar to Figure 7.8 except that the former has labeled tasks. In this case, the project pertains to the construction of a C programming language compiler.

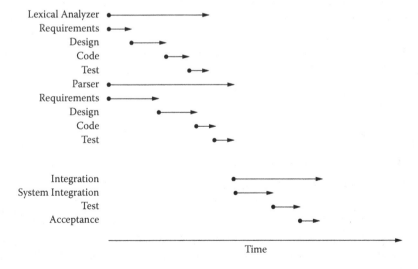

FIGURE 7.9
A line graph schedule.

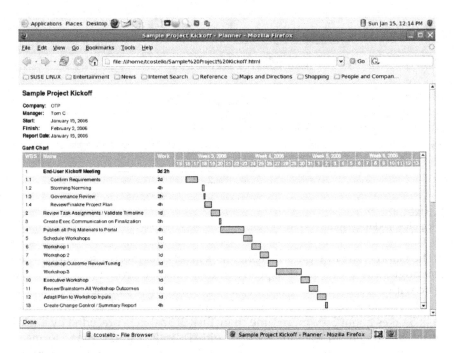

FIGURE 7.10
Screenshot of a Gantt chart built using Open Office.

Figure 7.10 is much more detailed than Figure 7.9; Figure 7.10 was created using an open-source tool called Open Office.[1] You can see that Figure 7.10 contains much more project information than the charts in Figures 7.8 and 7.9.

The level of refinement that you choose for a model—in this case, a project schedule—depends on several factors, including the number of persons involved in the project, the tools available, and the need for simplicity and convenience versus the need for a more detailed project plan.

7.4 Dealing with Figures

7.4.1 Callouts, Captioning, and Placement

There are some basic conventions in using figures in technical writing that you follow consistently. For example, every figure must have a caption, and the caption must have a unique number. The caption and number help in referencing the figure later. Figure 7.11a, for example, is incorrect because the figure is not captioned, and it is unclear how the figure relates to the surrounding text.

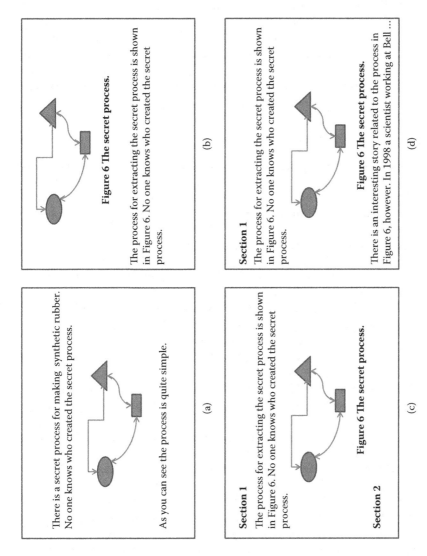

FIGURE 7.11
Some incorrect ways to caption figures (a–c) and the right way (d).

Any reference to the figure in the text is termed a "callout." Every figure must be called out. Even if the purpose of the figure seems obvious to you, you must call out that figure in the text.

Make sure that the figure appears after its first callout and not before. Figure 7.11b illustrates a first callout after the figure appears, which is incorrect. Figure 7.11d has the correct placement of the callout.

Don't end a section or chapter with a figure caption (as in Figure 7.11c). Instead, finish the chapter or section with a discussion pertaining to the figure after the figure caption (as in Figure 7.11d).

Finally, be prudent with the text in figure captions. A figure that requires an overly long caption may be too complex. If the figure needs elaboration, put the elaboration in the narrative, not in the caption.

7.4.2 Permissions for Figures

If you create an original figure, then you own the intellectual property rights (that is, the copyright) and you can use that figure however you like in your work. But if you plan to reuse a figure from another source, you must either cite the source and/or obtain proper permissions. Let me describe some of the nuances of figure permission rules.

If you would like to use a copyrighted figure, then you must obtain permission to use the figure (see Chapter 3). If you are going to redraw the figure and it is substantially different from the original, you do not need to obtain permission, but you do need to properly cite the original.

Figures 7.2 and 7.3 needed no permissions because I created them as originals. Because I paid the architect for his work, I own the copyrights for Figures 7.4, 7.5, and 7.6. I took the photograph in Figure 7.7, so I also own that copyright. Figure 7.1 needed to be cited because although I drew the figure, it displayed data from another source (cited in the caption). I needed to obtain permission to use Figure 3.5 because it is a replica of copyrighted material. The caption text was provided by the copyright holder, the IEEE.

Figure 4.4 is a special case. This figure appeared in a conference paper that was published in a proceedings document published by SPIE Press. However, the copyright transfer agreement with SPIE Press gives authors unlimited rights to use such intellectual property without needing further permission;[2] therefore, I only cite the source of the image.

Many images available on the Web are in the "public domain," meaning that their copyright has expired can be used freely. The rules about public domain are somewhat complicated and differ by country and by intellectual property type (for example, text, image, video, sound recording). Generally, however, an image that is 100 years or older than the passing of its creator will be in the public domain. But you should always confirm this fact before using it in a publication.

There are many figures in this book for which I owned the copyrights. But upon submitting the manuscript, I transferred these rights to the publisher, Taylor & Francis. In the future, I can use these figures in any other book

I write for Taylor & Francis without asking for permission—they already own the rights. If I wish to use these figures in some work for some other publisher, I will need to obtain permission from Taylor & Francis.

7.5 Dealing with Tables

Tables are an effective means of displaying complex data and textual information. You have seen many examples of tables throughout this book, some quite simple and others more complex.

You can use tables in unconventional ways. For example, Table 7.1 uses shading and numbers to generate a test image for a filtering algorithm.

TABLE 7.1

A Table Used to Generate an Image

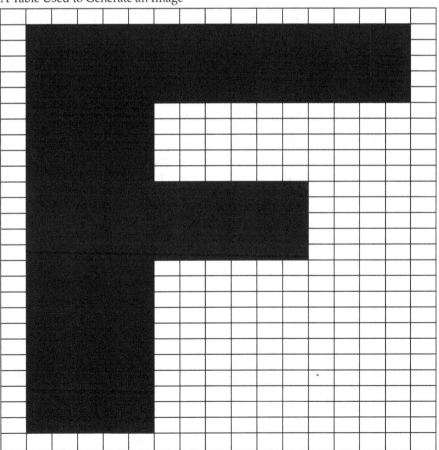

Table 7.2 shows another unconventional use of a table. Here, I used Microsoft Excel to create a spreadsheet that implements some Boolean logic to conduct a consistency check on a set of requirements. The rightmost five columns represent Boolean logic for five different requirements, and I used the Excel Boolean math functions to compute the values in these columns.

Aside from creative uses, I suggest that you keep tables as simple as possible. For example, multiple nested columns can be confusing. Take a look at Table 7.3, involving two years of harvest data from my vegetable garden. There is a lot of information in this table, but it would take a great deal of narrative to fully explicate the data. It might be better to describe highlights

TABLE 7.2

A Truth Table

P	Q	R	S	¬R	¬P	P <=>Q	Q=>R	¬RvS	¬P⇒S
TRUE	TRUE	TRUE	TRUE	FALSE	FALSE	TRUE	TRUE	TRUE	TRUE
TRUE	TRUE	TRUE	FALSE	FALSE	FALSE	TRUE	TRUE	FALSE	TRUE
TRUE	TRUE	FALSE	TRUE	TRUE	FALSE	TRUE	FALSE	TRUE	TRUE
TRUE	TRUE	FALSE	FALSE	TRUE	FALSE	TRUE	FALSE	TRUE	TRUE
TRUE	FALSE	TRUE	TRUE	FALSE	FALSE	FALSE	TRUE	TRUE	TRUE
TRUE	FALSE	TRUE	FALSE	FALSE	FALSE	FALSE	TRUE	FALSE	TRUE
TRUE	FALSE	FALSE	TRUE	TRUE	FALSE	FALSE	TRUE	TRUE	TRUE
TRUE	FALSE	FALSE	FALSE	TRUE	FALSE	FALSE	TRUE	TRUE	TRUE
FALSE	TRUE	TRUE	TRUE	FALSE	TRUE	FALSE	TRUE	TRUE	TRUE
FALSE	TRUE	TRUE	FALSE	FALSE	TRUE	FALSE	TRUE	FALSE	FALSE
FALSE	TRUE	FALSE	TRUE	TRUE	TRUE	FALSE	FALSE	TRUE	TRUE
FALSE	TRUE	FALSE	FALSE	TRUE	TRUE	FALSE	FALSE	TRUE	FALSE
FALSE	FALSE	TRUE	TRUE	FALSE	TRUE	TRUE	TRUE	TRUE	TRUE
FALSE	FALSE	TRUE	FALSE	FALSE	TRUE	TRUE	TRUE	FALSE	FALSE
FALSE	FALSE	FALSE	TRUE	TRUE	TRUE	TRUE	TRUE	TRUE	TRUE
FALSE	FALSE	FALSE	FALSE	TRUE	TRUE	TRUE	TRUE	TRUE	FALSE

TABLE 7.3

A Confusing Table

	Year Harvested					
	2009			2010		
		Yield			Yield	
Vegetable	Fertilizer Used	(lbs)	(bushels)	Fertilizer Used	(lbs)	(bushels)
Tomatoes	Super Grow	350	10	Super Grow	225	8
Peppers	Magic Yield	75	4	Wonderama	90	6
Eggplant	Wonderama	90	4	Magic Yield	85	4
Squash	Super Grow	100	2	Wonderama	110	3

from this table in words. For example, "the tomato yield in 2010 due to Super Grow was lower than in 2009." Alternatively, breaking up Table 7.3 into two tables—one for 2009 and one for 2010—might make the information easier to understand.

You should also avoid excessive use of tables—too many tables with little explanation can lead to dry writing. Finally, the conventions for permissions, captioning, numbering, and calling out tables are the same as for figures.

7.6 Dealing with Equations

A mathematical equation is just another model for something. Consider one of the most famous equations:

$$E = mc^2,$$

in which Einstein's model to explain the relationship between energy and matter in the universe. Another famous set of equations, Maxwell's, describes the relationship between electricity and magnetism (see Exercise 7.4).

Equations do not need to be cited unless they are new or not well known. The conventions for captioning, numbering, and calling out equations are the same as for figures with one exception. Not every equation must be numbered; you only number equations that will be called out. That is, if an equation is some intermediate calculation culminating in a final result equation, or an example of a previously numbered equation, then only the final result equation is numbered.

Here is an example of equation numbering from one of my unpublished works:

> The Fundamental Principle of Counting, or Fundamental Counting Principle (FCP), is an important concept in probability theory, as well in a wide range of areas that depend on probability theory such as modeling and simulation. The FCP is based on the notion of an experiment in which some activity leads to one or more outcomes. Typical "experiments" discussed in probability textbooks include flipping a coin (or coins), rolling a die (or dice), and drawing one or more cards from a standard deck of playing cards. But these pedestrian examples are only metaphors for real phenomena. In any case, here is the start of a rendition of the Fundamental Principle of Counting:

Suppose we are conducting an experiment (a process of some kind) incorporating k steps (stages). At each step i, there are n_i outcomes. Then the totality of the experiment will have

$$\prod_{i=1}^{k} i \tag{7.1}$$

possible outcomes.

To illustrate the use of the FCP, consider passwords consisting of exactly 8 characters (numbers and letters, case-insensitive) where the first must be a letter. How many possible passwords can be formed? By Equation 1 there are

$$26 \times 36 \times 36 \times 36 \times 36 \times 36 \times 36 \times 36 = 26 \times 36^7$$
$$= 2,037,468,266,496 \tag{7.2}$$

different passwords possible.

Of course, understanding this example requires that you know that the "Π" symbol represents multiplication over the range described by $i = 1$ through n. That is, Equation 1 is equivalent to $1{\cdot}2{\cdot}3{\cdot} \cdots n$ multiplied together. The point here is that if you are going to use equations and symbolic values, these must be defined prior to first usage. Also notice that the second equation is not numbered because it is not going to be referenced later in the writing.

Here is another example excerpted[3] from [Sinha and Laplante 2004]:

The set

$$\Theta \equiv \left\{ X_1, X_2, \cdots, X_n, Y_1, Y_2, \cdots, Y_n \right\} \tag{7.3}$$

is a chain with X_1 and Y_n as the minimal and maximal elements, respectively. We can write the multi-set Θ using the standard multi-set terminology as

$$\Theta = \left\{ P_1 \big|_{\lambda_1}, P_2 \big|_{\lambda_2}, \cdots, P_m \big|_{\lambda_m} \right\} \tag{7.4}$$

with

$$\lambda_i \geq 1,$$
$$\lambda_1 + \lambda_2 + \cdots + \lambda_m = 2n,$$
$$2n \geq m, \tag{7.5}$$
$$P_i \neq P_j \text{ for } i \neq j.$$

λ_i denotes the number of time s set P_i appears in the multi-set Θ. We also let $P_1 = X_1$ and $P_m = Y_n$, the minimal and maximal elements, respectively.

The rules for equation numbering are not inviolate—there are no "equation police." Just be consistent in your numbering scheme. For example, if you need to refer to only one or two equations in a sequence of equations, such as in a derivation, then number all the equations that are referenced. In the example, we see that only Equations 4 and 5 are numbered.

Typesetting equations can be both rewarding and extremely frustrating, particularly if you do not have the right tools. Good tools exist for typesetting all kinds of mathematical, electrical, and chemical equations.

7.6.1 Using Microsoft Equation Editor

Equation Editor is a proprietary equation typesetting tool that accompanies some versions of Microsoft's Office suite. Equation Editor was retired by Microsoft in January of 2018 but continues to be supported by third party vendors. Equation Editor, which must be configured as an option upon installation of Office, is compatible with Microsoft products including Word, Excel, and PowerPoint. Equation Editor is a graphical typesetting system: Through point-and-click, you build up the equation in line. All equations in this book were typeset with Equation Editor except as otherwise noted.

Equations typeset with Equation Editor are generally not compatible with document types that are incompatible with Microsoft Word, particularly document processing software for the Linux operating system. There are various free and commercial converters between typesetting languages but I have had trouble with some of these (such as system crashes and incorrect or incomplete conversion), so be wary of using equation typesetting converters.

7.6.2 Using MathType

MathType, developed by Design Science, is a more powerful and platform independent version of Equation Editor. MathType is easier to use than Equation Editor, particularly for building very complicated equations. Unlike Equation Editor, equations typeset with MathType are compatible with a wide range of non-Microsoft document types and operating systems. Therefore, if you expect to have many equations in your technical writing, I suggest that you purchase MathType.

7.6.3 Using LaTeX

LaTeX is a typesetting language that can handle very complex mathematical equations and tables [Griffiths 1997; Lamport 1994]. LaTeX[4] is a generic name and proprietary and open source versions are also available. LaTeX is not graphical in nature. Instead, it uses plaintext input and a rather cryptic markup language to achieve its result. Although it is cumbersome to use, LaTeX is actually an easier-to-use version of the TeX typesetting language developed by Donald Knuth in the 1980s, which in turn was based on the Unix mathematics typesetting language *eqn* developed at Bell Laboratories in the early 1970s [Kernighan 1975]. There are a number of proprietary and open-source implementations of this software [LaTeX].

I typeset my doctoral dissertation using LaTeX more than twenty years ago. My dissertation spanned more than 200 pages and contained many complex equations. It took me nearly as much time to typeset the equations as it did to conduct the research.

An example of a LaTeX source document is shown in Figure 7.12. The resulting typeset document excerpt is shown in Figure 7.13. A more elaborate example of an equation in LaTeX is shown in Figure 7.14, and the result of typesetting the LaTeX source in Figure 7.14 is shown in Figure 7.15. Notice in Figure 7.15 that only the last, summary equation is numbered because it was called out later in the paper but the other equations were not.

The equations shown in Figure 7.15 were particularly challenging to typeset. Moreover, the source conference paper [Sinha and Laplante 2002] was eventually expanded into a journal-length paper [Sinha and Laplante 2004]. But the journal only accepted equations typeset using Equation Editor or MathType. I attempted to use equation-converting software but with unacceptable results. Eventually, I had to re-typeset all the equations in MathType, which was an enormous effort.

```
Let ${(\rm (\bf A)}$ be an object and let ${(\rm (\bf \vec (x)}} \in (\rm (\bf
A)}$ be a pixel point. Let us assume that the range of image positioning
error is $\pm \Delta $. The error range can be zero, less than one pixel, or
one or more pixels.

For each positioning error $\delta \in \left[ ( - \Delta ,\Delta } \right]$,
one can determine the upper and lower approximations of the object ${(\rm (\bf
A)}$. This collection of all upper and lower approximations will be
finite\footnote{ The collection will be finite so long as $\Delta $ is
finite. Very few positioning errors will lead to unique pairs of upper and
lower sets. See Fig.~\ref{fig6}and the worked example at the end of the paper.}
and let us denote them as (for some $n \ge 1)$


\[
\left\{ (\left\langle ((\rm (\bf L)}_i (\rm (\bf ,U)}_i } \right\rangle :i =
1,2, \cdots ,n} \right\}).
\]

Why do we wish to consider such a collection? Intuitively speaking, the
pixels in the set $\bigcap\limits_{i = 1)^n ((\rm (\bf L)}_i } $ are almost
definitely going to be present in the image no matter what the positioning
error. Similarly, pixels that do not belong to the set $\bigcup\limits_{i =
1)^n ((\rm (\bf U)}_i } $ are not going to be present in the image. For all
other pixels we need to determine a likelihood that they will appear in the
image. We will do this by inspecting all the lower and upper approximations.

Define the following sets:
```

FIGURE 7.12
LaTeX source for Figure 7.13. (From Sinha, D. and Laplante, P., *Proc. Rough Sets and Current Trends in Computing Conference*, PSU Great Valley, October 2002, pp. 610–620.)

Let $\mathbf{A}$ be an object and let $x \in \mathbf{A}$ be a pixel point. Let us assume that the range of image positioning error is $\pm \Delta$. The error range can be zero, less than one pixel, or one or more pixels.

For each positioning error $\delta \in [\ \Delta, \Delta]$, one can determine the upper and lower approximations of the object $\mathbf{A}$. This collection of all upper and lower approximations will be finite[3] and let us denote them as (for some $n \geq 1$)

$$\{\langle \mathbf{L}_i, \mathbf{U}_i \rangle : i = 1, 2, \cdots, n\}.$$

Why do we wish to consider such a collection? Intuitively speaking, the pixels in the set $\bigcap_{i=1}^{n} \mathbf{L}_i$ are almost definitely going to be present in the image no matter what the positioning error. Similarly, pixels that do not belong to the set $\bigcup_{i=1}^{n} \mathbf{U}_i$ are not going to be present in the image. For all other pixels we need to determine a likelihood that they will appear in the image. We will do this by inspecting all the lower and upper approximations.

Define the following sets:

---

[3] The collection will be finite so long as Δ is finite. Very few positioning errors will lead to unique pairs of upper and lower sets. See Fig. 6and the worked example at the end of the paper.

FIGURE 7.13
Typeset documents from the source shown in Figure 7.12. (From Sinha, D. and Laplante, P., *Proc. Rough Sets and Current Trends in Computing Conference*, PSU Great Valley, October 2002, pp. 610–620.)

LaTeX has associated tools such as BibTeX, which helps you organize both bibliographies and relative citations so that if you move equations around, the numbering is maintained properly. Some publishers, for example for conferences, will only accept papers typeset using LaTeX.

Today I use Microsoft Equation Editor and MathType most of the time. But because it is a free tool and is platform independent, many authors, especially mathematicians and physicists, some version of LaTeX.

7.6.4 Vignette: Typesetting Books

I had significant experience using LaTeX from typesetting several complex papers and my dissertation and, at the time (around 1995), I considered myself an expert. So when a publisher required a camera-ready[5] book, I decided to typeset it using LaTeX, I thought it would be no problem. Boy, was I wrong. I really struggled with getting the margins, pagination, and flow to meet the publishing requirements—at the time, anyway, LaTeX was not a helpful tool to control layout. The worst part was that I messed up the automatic indexing. So, in early editions of the book, using Unix [Laplante and Martin 1998], the index lists the first page as 12 not 1. This problem occurred because I didn't get the page counting mechanism to discount the front matter pages (which are numbered with Roman numerals) and start the first page at 1.

```
\begin{eqnarray*}
{\rm {\bf X}}_1 = \bigcap\limits_{i = 1}^n {{\rm {\bf L}}_i } ,
\quad
{\rm {\bf X}}_2 = \bigcup\limits_{i = 1}^n
{\bigcap\limits_{\begin{array}{l}
 j = 1 \\
 i \ne j \\
 \end{array}}^n {{\rm {\bf L}}_j } } ,
\quad
{\rm {\bf X}}_3 = \bigcup\limits_{i = 1}^n
{\bigcup\limits_{\begin{array}{l}
 j = 1 \\
 j \ne i \\
 \end{array}}^n (\bigcap\limits_{\begin{array}{l}
 k = 1 \\
 k \ne i \\
 k \ne j \\
 \end{array}}^n {{\rm {\bf L}}_k } } } , \quad \ldots \ldots ,\\
\quad
{\rm {\bf X}}_n = \bigcup\limits_{i = 1}^n {{\rm {\bf L}}_i } ,
\end{eqnarray*}

\begin{eqnarray}
\label{eq3}
\nonumber
{\rm {\bf Y}}_1 = \bigcap\limits_{i = 1}^n {{\rm {\bf U}}_i } ,
\quad
{\rm {\bf Y}}_2 = \bigcup\limits_{i = 1}^n
{\bigcap\limits_{\begin{array}{l}
 j = 1 \\
 i \ne j \\
 \end{array}}^n {{\rm {\bf U}}_j } } ,
\quad
{\rm {\bf Y}}_3 = \bigcup\limits_{i = 1}^n
{\bigcup\limits_{\begin{array}{l}
 j = 1 \\
 j \ne i \\
 \end{array}}^n {{\rm {\bf U}}_j }{\bigcap\limits_{\begin{array}{l}
 k = 1 \\
 k \ne i \\
 k \ne j \\
 \end{array}}^n {{\rm {\bf U}}_k } } } , \quad \ldots \ldots ,\\
\quad
{\rm {\bf Y}}_n = \bigcup\limits_{i = 1}^n {{\rm {\bf U}}_i } .
\end{eqnarray}
```

FIGURE 7.14

A more complex LaTeX source file. (From Sinha, D. and Laplante, P., *Proc. Rough Sets and Current Trends in Computing Conference*, PSU Great Valley, October 2002, pp. 610–620.)

$$\mathbf{X}_1 = \bigcap_{i=1}^{n} \mathbf{L}_i, \quad \mathbf{X}_2 = \bigcup_{i=1}^{n} \bigcap_{\substack{j=1 \\ i \neq j}}^{n} \mathbf{L}_j, \quad \mathbf{X}_3 = \bigcup_{i=1}^{n} \bigcup_{\substack{j=1 \\ j \neq i}}^{n} \bigcap_{\substack{k=1 \\ k \neq i \\ k \neq j}}^{n} \mathbf{L}_k, \quad \ldots \ldots,$$

$$\mathbf{X}_n = \bigcup_{i=1}^{n} \mathbf{L}_i,$$

$$\mathbf{Y}_1 = \bigcap_{i=1}^{n} \mathbf{U}_i, \quad \mathbf{Y}_2 = \bigcup_{i=1}^{n} \bigcap_{\substack{j=1 \\ i \neq j}}^{n} \mathbf{U}_j, \quad \mathbf{Y}_3 = \bigcup_{i=1}^{n} \bigcup_{\substack{j=1 \\ j \neq i}}^{n} \bigcap_{\substack{k=1 \\ k \neq i \\ k \neq j}}^{n} \mathbf{U}_k, \quad \ldots \ldots,$$

$$\mathbf{Y}_n = \bigcup_{i=1}^{n} \mathbf{U}_i. \quad (3)$$

FIGURE 7.15
Result of LaTeX typesetting for the code shown in Figure 7.14. (From Sinha, D. and Laplante, P., *Proc. Rough Sets and Current Trends in Computing Conference*, PSU Great Valley, October 2002, pp. 610–620.)

I managed to discover and correct the problem before too many copies of the book were produced and delivered.

I thought I had learned my lesson about producing camera-ready copy, so when a different publisher asked me to produce a second camera-ready book, I balked. But the publisher convinced me that Microsoft Word had good support for automatic generation of the bibliography, table of contents, and index and would be well suited to producing the book. What I didn't realize is that the touted support is semi-automatic—you have to diligently tag many items with metadata or these automatic features don't work well. Word can also by tricky when dealing with pagination, page breaks, headers, margins, spacing and other issues related to layout. Again, I was not the expert I thought I was, and producing this computer architecture book [Gilreath and Laplante 2003] was not as easy as I had hoped.

I was very disappointed with the first results of my camera-ready book, and had to invest many hours to improve the result. I think the lesson here is to be realistic about the large amount of work needed to produce camera-ready copy.

7.7 Dealing with Dynamic Content

Word processing systems, such as Microsoft Word, allow you to embed certain dynamic objects within the electronic document. For example, the truth table shown in Table 7.2 is dynamic in that if you change some of the truth values in the leftmost columns, some of the values in the five rightmost columns may change. Occasionally, some kind of dynamic content—a moving picture or graphic—might be the most effective way of conveying information. Of course, you can't have moving pictures in a printed document. But if the document is stored in some electronic media, such as on a disk or a website, then dynamic content could be embedded within the document. For example, it is common to incorporate video in webpages. But for a printed document, static representation schemes are necessary.

One approach is to provide a link to a website hosting the dynamic content and reference that site in your document. This technique was used in a paper I cowrote a few years ago. The paper described a technique for controlling airplane propeller pitch for small, single-engine airplanes in a novel way [Nash and Laplante 2004]. We built a computer simulation for the control system and validated the proposed system. However, the static screenshots in the published paper just didn't convey the desired impression. So, we used a link to a simulation video hosted on my website. Here is an excerpt from that paper:[6]

> A screen layout was developed. It is presented as Figure 7.16. The aircraft can be flown according to a script (Autopilot). Script parameters are supplied through a series of six numeric boxes. If the engine is on in this mode the aircraft takes off and flies the runway heading until it reaches its Turning Altitude. This value is obtained from the first numeric box.
>
> The aircraft continues its climb and turns at standard rate to its initial heading (from a numeric box). Upon reaching the initial heading, the climb continues to cruising altitude (also obtained from a numeric box). Length of cruising flight is supplied by the numeric box entitled, Flight Time. After this time has elapsed the aircraft begins its descent to an approach altitude. Accomplishing the descent, it turns to the runway heading. Descent to landing follows.
>
> A set of multimedia files showing the simulator in action (Microsoft Media Player compatible file) can be found at https://phil.laplante.io /resources.php. The simulation consists of four files. A better understanding of these files can be attained by the reader referring to the screen layout in Fig. 4 prior to viewing a particular video. [Nash 2004]

You can watch these videos by visiting my personal website at https://phil .laplante.io/resources.php.

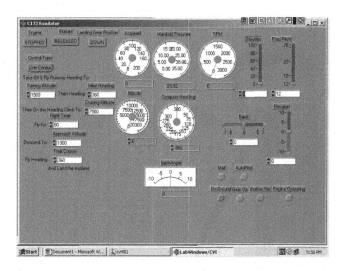

FIGURE 7.16
Screen layout of system emulator. (*Source:* From Nash, J. and Laplante, P., *Journal of Aerospace Computing, Information, and Communication*, 1(1), 198–212, 2004. Reprinted with permission of the American Institute of Aeronautics and Astronautics.)

7.7.1 Vignette: The Minard Map

Charles Minard's 1861 map depicting Napoleon's invasion of Russia in 1812 (Figure 7.17) has been called the "the best statistical graphic ever drawn" [Tufte 2001]. The graphic depicts an initial 442,000 troops as a thick line superimposed on a map of region. As your eye follows the line of march you can see the dwindling number of troops due to the terrible winter conditions (temperatures are shown at the bottom) and battle (represented by the named sites).

The caption, written in nineteenth century French, translated loosely, states that the image indicates the number of men present by the thickness of the lines at a rate of one millimeter per ten thousand men and that the red (light gray in this version) coloring signifies the men who entered Russia, and the black those who escaped (only 10,000 returned). The caption further lists the sources for the data represented.

Imagine the effort to describe the information contained in the picture in words only. For further discussion of this case and two other famously effective graphics see *The Economist* [2007].

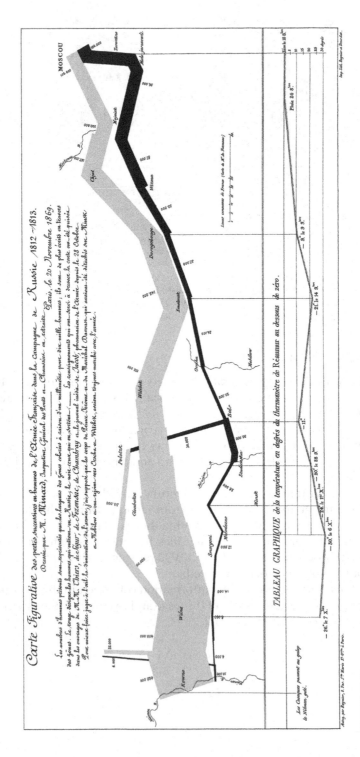

FIGURE 7.17

Charles Minard's 1861 chart of Napolean's 1812 Russian Campaign. (*Source:* https://en.wikipedia.org/wiki/File:Minard.png in the public domain).

7.8 Exercises

7.1 Locate a figure in this book that is reprinted with permission.

7.2 Locate a figure in this book that is cited but for which no permission was obtained. How does this figure differ from the one you located in Exercise 7.1?

7.3 Find "Bernoulli's Equation" on the Web and typeset it in a document using any equation preparation system available to you.

7.4 Find "Maxwell's Equations" on the Web and typeset these in a document using any equation preparation system available to you.

7.5 Try to explain Figure 7.2 in words using as few words as possible. How many words did it take?

7.6 Draw a figure of any type depicting how you engage in your favorite hobby.

7.7 Research then describe the "Ideal Gas Equation", that is, $PV = nrT$, in words only.

7.8 The following is one form of an ancient proverb (it exists in many cultures and languages):

For lack of a nail a horseshoe was lost,

for lack of a horseshoe a horse was lost,

for lack of a horse a rider was lost,

for lack of a rider a message was lost,

for lack of a message an army was lost,

for lack of an army a battle was lost,

for lack of a battle a war was lost,

for lack of a war a kingdom was lost,

and all for want of a nail.

Convert this into some form of graphic (picture, equation, chart, etc.).

7.9 Research William Playfair's famous 1821 chart on wages and write a 200-word summary essay.

7.10 Research Nightingale's Coxcomb (sometimes called "Nightingale's Rose") and write a 200-word summary essay.

Endnotes

1. Thanks to my friend Tom Costello of Upstreme Inc. for introducing me to this program and furnishing this example.

2. Here is the legal language: "Authors retain the right to prepare derivative publications based on their own paper, including books or book chapters and magazine articles, provided that publication of a derivative work occurs subsequent to the official date of publication by SPIE" [SPIE 2010].

3. This excerpt offers another example of a different kind of permissions. The copyright holder, Elsevier, provides the author with "the right to prepare other derivative works, to extend the journal article into book-length form, or to otherwise re-use portions or excerpts in other works, with full acknowledgement of its original publication in the journal." (https://www.elsevier.com/__data/assets/pdf _file/0006/98619/Sample-P-copyright-2.pdf, last accessed 12/29/2017.)

4. LaTeX is not an acronym. The "La" stands for Leslie "Lamport," the inventor, and "TeX" is a play on the prefix "Techni."

5. "Camera ready" means that the material provided is ready to be printed without further editing or formatting.

6. Reprinted from Nash, J. and Laplante, P. *Journal of Aerospace Computing, Information, and Communication,* 1(1), 198–212, 2004. With permission of the American Institute of Aeronautics and Astronautics.

References

Gilreath, W. F. and Laplante, P. A., *Computer Architecture: A Minimalist Perspective,* Springer US, 2003.

Griffiths, D. F., *Learning LaTeX,* Society for Industrial and Applied Mathematics, Philadelphia, PA, 1997.

Kernighan, B. W. and Cherry, L. L., A system for typesetting mathematics, *Communications of the ACM,* 18, 151–157, 1975.

Lamport, L., *LaTeX: A document Preparation System, Second Edition,* Addison-Wesley Professional, Reading, MA, 1994.

Laplante, P. A. and Martin, R., *Using UNIX,* West Publishing Co., New York, 1998.

LaTeX—A document preparation system, http://www.latex-project.org/, accessed December 29, 2017.

Nash, J. and Laplante, P., A real-time control system to control aircraft propeller pitch, *Journal of Aerospace Computing, Information, and Communication,* 1(1), 198–212, 2004.

Lee, P. S., West J. D., Howe B., Viziometrics: Analyzing visual information in the scientific literature. *IEEE Transactions on Big Data*, March 29, 2017.

Sinha, D. and Laplante, A., Handling spatial uncertainty in binary images: A rough sets approach, *Proc. Rough Sets and Current Trends in Computing Conference*, PSU Great Valley, October 2002, pp. 610–620.

Sinha, D. and Laplante, P. A., A rough set based approach to handling spatial uncertainty in binary images, *Engineering Application of AI*, 17(1), 97–110, 2004.

SPIE, Obtaining Permission to Use Previously Published Material from an SPIE Journal, 2010. SPIE website, http://spie.org/x1811.xml; accessed December 29, 2017.

Tufte, E., *The Visual Display of Quantitative Information*, Second Edition, Graphics Press, Cheshire, Connecticut 2001.

Worth a thousand words, *The Economist*, 385(8560), December 19, 2007, p. 74.

8

Publishing Your Work

8.1 Introduction

In Chapter 8 I discuss two major themes. The first relates to publishing your own work. Let's suppose that you have written a technical article or a book, and you believe that this writing is suitable for publication. You ask, "Where can I publish this work, and how do I get it published?" The second theme is a response to the hypothetical question of "Is it possible to make a living as a technical writer, and if so, how?" This chapter helps answer these questions.

I don't know why you are reading this book. Perhaps it was required for a course. Perhaps you acquired this book because you write as a hobby. Whatever the case, you are probably a technical professional or plan to be. Every technical professional has the potential to publish something relating to his (her) own work experience. I believe that it is in your best interest to consider publishing your work, whether for professional advancement, identity branding, personal satisfaction, or for supplemental income.

8.1.1 What Kinds of Work Can Be Published?

You can publish almost anything: basic and applied research, experience reports, surveys of the literature or a professional area, even opinion pieces. You can publish in various places, including conferences, books, magazines, newsletters, journals, newspapers, websites, and blogs. Where you publish depends on the type of material.

Selecting the right publishing venue is crucial because submitting a manuscript to the wrong venue can lead to bitter disappointment and delays in publication. Every writer knows the disappointment of rejection, but you can learn from that experience. Dealing with rejection is explored in Section 8.4.

8.1.2 Why Publish Your Work?

Even if you don't plan on making a living as a technical writer, and you only write as a part of your job as a scientist or technical professional, there are several reasons why you should consider publishing in a magazine, journal, or conference. First, having publications suggests that you are an "expert." Suppose you have maintained a vegetable garden in your backyard for twenty years and you think you have become quite proficient. But how do you know for sure that your "success" as a gardener is extraordinary? One way to benchmark your accomplishments is to write an article about some aspect of gardening that you think you have mastered, for example, growing asparagus. Then submit that article to a reputable gardening magazine so that it goes through a vetting or review process. If the article is accepted for publication, then you will have independent validation of your expertise in growing asparagus and, more importantly, you will have public evidence of your expertise—the published article. I noted in Chapter 3 how even one or two published works on an otherwise "generic" résumé can distinguish you from other job applicants. If you were applying for a job as an asparagus-farming technician, having a reviewed article in your portfolio would be most helpful. I remember that when I would interview for industrial positions, my list of publications would always draw acclaim.

Another reason to publish your work is simply for personal satisfaction and ego gratification. I still get a thrill when an article is accepted in a journal, magazine, or conference—even after having published 200 previous items.

A more noble reason for publishing your work is because you have some important discovery or insight that you want to share or simply because you wish to "give back" what you have learned as a service to others. These reasons may be the most important of all because they can motivate you to publish without the expectation of any personal reward.

Finally, you can try to publish your work to make money. But how much money can you make publishing technical articles? Can you make a living from technical writing? I will offer my opinion on these questions in the next section.

8.2 Making a Living as a Writer

Many companies hire professionals to write and edit various kinds of technical publications, including user manuals, installation instructions, company newsletters and brochures, advertising copy, and text for product packaging. This kind of work can be interesting and rewarding.

But what if you want to write your own material, that is, you want to pick and choose technical topics that interest you to write about, and you want to publish that writing for a living income? Can you really earn a livelihood this way? Of course, the answer to that question is "yes," and I know several people who make their living as freelance writers. Let's see how.

8.2.1 Freelance Writing

"Writing freelance" means that you work as an independent consultant and have to find and serve an ever-changing list of clients. Usually you will be required to write an article on some topic for a magazine, or to edit or fact check someone else's article. You may also be asked to serve as an editor and help find and coach other authors. You can find these assignments online, through personal and professional contacts, and later, through referrals.

I doubt you will get rich as a freelance writer because technical writing and editing is very labor intensive, and you are often juggling multiple deadlines for projects that conflict. As an independent contractor, the amount of money you make depends on the number of hours you can work, but you will not be paid an hourly wage; rather, you are usually paid a negotiated fee for each project. You can probably make $50,000 to $60,000 annually as a freelance technical writer if you really hustle, but you have the benefits of being able to choose projects, work at home, and manage your own work schedule.

If you keep your regular job and choose to write in your spare time, what kind of money can you expect? First, recognize that most scholarly magazines, journals, and conferences do not pay for articles. Often, publishing in these venues costs you money in terms of travel (to the conference) or charges to the author for publishing the work. You may hear the term "page charges," referring to costs that the author must pay to have his work published in some scholarly journals; these charges are based on the length of the work in terms of dollars per page. On the other hand, some trade magazines will pay $200 to $500 for a published article. My fish tank article yielded a few hundred dollars (I can't recall the exact amount), and it was one of only two articles for which I was paid.

8.2.2 Writing Technical Books

There are essentially four types of target technical books: basic introductory (e.g., "dummies") books, classroom textbooks, advanced monographs, and trade books. Introductory books are too basic for classroom use, but serve hobbyists, young people, and continuous learners.[1] Introductory books have the largest potential sales market, but it is very difficult to capture a large market share.

Classroom textbooks are targeted for undergraduate and graduate courses at colleges and universities. Depending on the size of the market, it is possible to sell tens of thousands of copies of such a text over a few years. But only a few texts in any field realize this sales potential—for example, the top-selling first-year calculus, chemistry, or biology textbooks. For most other technical books, sales of 2000 or 3000 copies would be considered successful.

Advanced monographs, or simply "monographs," serve a tiny group of researchers and students in a very specific field, for example, advanced graph theory. These books are called "monographs" because they cover a single focused topic, and are often written by a single author. Books of this type usually have a very high purchase price, and the publisher expects to sell only a few hundred copies. In some cases, the author is required to deliver camera-ready copy; that is, the author must take care of editing and typesetting so that the book is ready to be copied and bound. Because the book anticipates a low unit sales volume, the publisher cannot make a large upfront investment in the book's production. I have cowritten one book of this type, a highly esoteric one on minimalist computer architecture [Gilreath and Laplante 2003]. Some consider advanced monographs as vanity-type books, but they provide an important mechanism for professors to archive and propagate advanced graduate-level material.

"Trade" or "professional" books are intended for practitioners. *Advanced AutoCAD* and *Introduction to Java* are typical titles. Trade books are intended to be sold at high volume for a relatively low price.

Occasionally, a book will serve both the classroom text and professional markets, to the delight of both publisher and author. *Real-Time Systems Design and Analysis* is one such example [Laplante 2004].

8.2.3 Getting Rich Writing Books

Don't think you can get rich writing technical books. Technical books rarely yield a large financial return compared with the investment of your time and the publisher's time and money. Author royalties are typically in the range of 5% to 20% of adjusted net sales less returns. The royalty rate depends on many factors, including bulk buying discounts, overseas sales, softcover versus hardcover, etc. Often, the royalty rates are graduated.

For example, here is a simplified version of the royalty description section from one of my book contracts:

(1) From sales or licenses of the original edition whether in hardcover, paperback or other medium (the "regular edition") in the United States, its possessions and territories, and Canada:

10% on the first 1000 copies sold;

12.5% on the next 1000 copies sold;

15% on all additional copies sold.

(2) From sales or licenses of the regular edition elsewhere:

10% on the first 1000 copies sold;

12.5% on the next 1000 copies sold;

15% on all additional copies sold.

(3) From sales or licenses of the Work and materials from the Work in electronic form, whether directly by the Publisher or indirectly through or with others:

5%

(4) From sales of lower priced paperback editions throughout the world:

5%

(5) From sales of any edition through direct-to-consumer marketing (including, for example, direct mail, but not including sales made via the Publisher's website):

5%

(6) From sales of copies of the Work produced "on demand", when it is not feasible to maintain a normal inventory:

5%

(7) From sales of the Work at discounts of 50% or more from list price or sold in bulk for premium or promotional use, or special sales outside the ordinary domestic channels of trade:

50% of the applicable royalty rate

(8) From sales or licenses by the Publisher to third parties of the following subsidiary rights in the Work: book club, reprint, serial rights, foreign language translation rights, audio and video adaptation rights, and in any media, permission grants for quotations of short excerpts and photocopies, after deduction of Publisher's out-of-pocket costs, if any, incurred in connection with such sales or licenses:

50%

(9) Should Publisher undertake, either alone or with others, the activities described in (8) above:

10%

(10) From sales or licenses of condensations, adaptations, and other derivative works not specified above:

10%

(11) From use of all or a part of the Work in conjunction collectively with other work(s), a fraction of the applicable royalty rate equal to the proportion that the part of the Work so used bears in relation to the entire collective work:

- pro rata

Note how many of the rates are dependent on crossing the magic 1000-book threshold.

Let's take an example. Suppose a technical book for classroom use has a list price of $100, which is fairly common for a university textbook.[2] A retailer will pay no more than $60 for that book—the price per copy will be lower for a large order. To simplify the math, let's say for every copy sold, the publisher gets $60 and I have a flat royalty rate of 15%. I earn $9 for every book sold. If 1000 books are sold over a few years (which is good for most types of books),

I earn $9000, before paying taxes on those earnings. Although I usually have three or four books paying royalties at any given time, I can't quit my regular job. And here is why.

I have had several books that have sold less than 1000 copies over their lifetime. Bootleg copies of my books exist (see Section 8.6.3), and I earn nothing from these. After a while, books are too old to be sold and earn royalties. And for every book that is returned to the publisher as unsalable, the author is debited for the corresponding royalty. That kind of arrangement usually doesn't add up to very much for the author. It is very unlikely that you can earn a living solely from writing technical books unless your books are wildly successful or you have the time, talent, and energy to write lots of them.

Aside from the personal gratification and modest remuneration, however, there is one other benefit to writing books—the establishment of credibility. If writing one or two articles on a subject in conference proceedings, a magazine, or a journal can establish your expertise, then writing a book really enhances your credibility. And so, the additional benefit is that, on the strength of your published book, you may be called upon, as I have been, to do consulting or to give a talk, or to serve as an expert in the area in which you have published.

If you are making a living as a technical consultant, then you should consider writing a book. There is no better "calling card" to a prospective client than a copy of your book. However, be wary here too—publishers do not give authors unlimited copies of their book for free. You will receive a few free copies at publication time (usually between five and ten), but after that you are required to buy your own book, usually at a 40% discount.

8.2.4 Why Are Technical Books So Expensive?

I have already noted that technical books, particularly for university classroom use, often cost $100 or more and also that I might receive $9 or less per copy (for this book, much less), depending on the kind of sale. You may conclude, then, that the publishers are ripping you (and me) off, but they are not.

There is a tremendous amount of money invested in the editing, review, layout, marketing, sales, production, and shipping of a technical book. For a typical technical book, the publisher's investment can be on the order of tens of thousands of dollars, and there is risk involved. This investment must be made before the first sale, and a period of two to four years or more may elapse before the first copy sells and the publisher and author see any revenue.

For example, although a book may cost the purchaser $100, the publisher sells the book to a retailer for around $60. But because of volume discounting, a "big" sale of a hundred books may result in the publisher receiving $40 or less per book. Then the author must be paid his royalty. Finally, subtract sales lost to resold books, used books, bootleg copies, and book rentals, which all represent increasing threats to the profits of publishers and authors. Taken together, a typical technical book usually needs to sell 1000 copies in order for the publisher to recoup its investment, and not all titles will sell 1000 copies.

So how can publishers make money if some of their books lose money? The answer lies in portfolio theory. Publishers know that they must sell a number of books at a loss in order to offer a comprehensive portfolio of technical books so that they can hope to continue to sell their profitable books. Let me illustrate the situation in a different way. An ice cream store may rely on profits from its three biggest-selling ice cream flavors—chocolate, vanilla, and strawberry—to pay its costs and make a reasonable profit. But can an ice cream store remain in existence selling only these three flavors? No, the ice cream store must sell thirty or more other flavors, all of them at a loss or break even, to be considered an enticing ice cream store that we want to patronize.

My discussion above is somewhat simplified, of course. But the situation is really as I describe it. So, don't be angry at the publisher or author when you have to spend $100 for a technical book.

8.2.5 Vignette: A Writing Failure[3]

When I first started teaching in the early 1990s, I needed a list of important papers in the field of computer science for a course I was preparing. When I couldn't find a satisfactory list, I decided to compile my own list by surveying other professors of computer science. I was impressed with the collection of papers, many written by computer science pioneers. The resulting list represented what I believed to be a set of "great papers" in computer science.

I obtained copies of these papers, which was rather difficult because there were few digital libraries from which to draw sources. Many of the papers had to be ordered directly from the publisher via surface mail.

I published the list of papers with some annotations for instructors in 1994 [Laplante 1994]. I also interested a publisher in a collected volume of the reprinted papers accompanied by retrospective introductions by any of the original authors that I could find. The book was to be called *Great Papers in Computer Science*, the name I gave my original survey list.

The *Great Papers* book, while a great idea, was trouble from the start. While I wanted to simply photocopy the original papers, the publisher insisted on manually retyping them. The task turned out to be a disaster because many of the symbols used in the early papers were unavailable on the word processors of the time, and the retyping introduced a very large number of errors, many of which escaped review. Cost overruns prevented additional rounds of proofreading and the book was rushed to market.

Several of the contributing authors contacted me to express their dissatisfaction with the quality of the book, although a few praised the effort. But I was quite disappointed with the result and how it might affect my reputation.

The book is now long out of print, although you can occasionally find it for sale online. As a consolation to me, the book is cited frequently because many of the papers are still hard to find. I also reacquired the copyright to the book in the hope that I could eventually republish a high-quality version and redeem myself.

8.3 The Review Process

All work that is submitted for publishing consideration undergoes some form of review. The review may be by peers, that is, by other experts in the field, by the editorial staff of the publication for clarity and correctness, and in most cases, by both groups. An exception to this rule is if you have been invited to submit something because of your status as a leading expert in some field. But don't be misled: Even Nobel Prize winning scientists will have their work reviewed in the conventional way when they submit a paper for consideration in a scholarly journal.

Let me outline the review process for submitted works for publishing consideration. This review process is essentially the same for papers submitted to journals, high-quality technical magazines, conferences, and book proposals. Less rigorous reviewing might be in place for newsletters and informal magazines, but some kind of quality control for content will be in place.

8.3.1 Administrative Rejection

As soon as a manuscript is received by a publisher; by the editor of a journal, magazine, or newsletter, or by a conference program committee chair, it is evaluated to see if the manuscript can be put into the formal review process. A formal review takes a great deal of time and effort by volunteer (and sometimes paid) reviewers, so there is an incentive to determine if a manuscript can be rejected without a full review. Such a rejection is called an "administrative rejection," which can happen for one or more of the following reasons:

- The submission is not within the stated scope of the journal, conference, magazine, or book publisher.
- The quality of writing is too poor for a proper review to take place.
- The length of the manuscript is out of specification (too short or too long).
- The work described in the manuscript is of obviously poor quality.

Editors and publishers do not make administrative rejection decisions lightly. They understand that even a badly written manuscript can have technical merit. Therefore, they will usually return the manuscript to the authors with advice for improvement and invite them to either resubmit the manuscript or direct them to a more appropriate venue.

8.3.2 Review Flow

If a manuscript is not administratively rejected, then it will enter the formal review process. A simplified representation of the review process for a paper is illustrated in Figure 8.1.

The review process starts with submission of the manuscript either via e-mail or website. Most periodical publishers now use Web-based submission and review tracking systems.

Once the manuscript has been submitted, and when the editor-in-chief (EIC) or managing editor has decided not to administratively reject it for the reasons previously mentioned, then the manuscript goes into review. The editor will choose two to five experts in the subject area of the submitted manuscript to review it. Depending on the type of publication and the size and complexity of the manuscript, the reviews may take 2 to 18 months to complete. Newsletters, magazines, and conferences sometimes boast of review times of two weeks or less—and occasionally, these expectations are met.

The reviewers will return their recommendations to the EIC. The recommendation can be accept the manuscript, accept with minor changes, accept with major changes, rewrite the manuscript and resubmit, or reject. The EIC takes these recommendations into consideration and arrives at a decision, which is communicated to the author. In the case of accept with major or minor changes, the author may make the recommended changes and resubmit the manuscript. The EIC can decide whether the manuscript should be reviewed again or sent directly into production where it will be copyedited and formatted.

I have simplified the description of the process greatly and there are many variations that can occur. Sometimes the EIC does not get enough reviews or the reviews are so varied that the EIC has difficulty making a decision. In these cases, some negotiation by the author with the EIC may be possible in order to get the manuscript accepted with minor or major revision.

The reviews that come back to the EIC and eventually to the authors can be helpful and filled with constructive criticism. But the reviews can also be brutally honest, hurtful, sarcastic, insulting, or misinformed. Learning how to deal with negative feedback takes time, and you must develop the ability to take harsh criticism without getting angry. These skills can be attained by working with an experienced mentor.

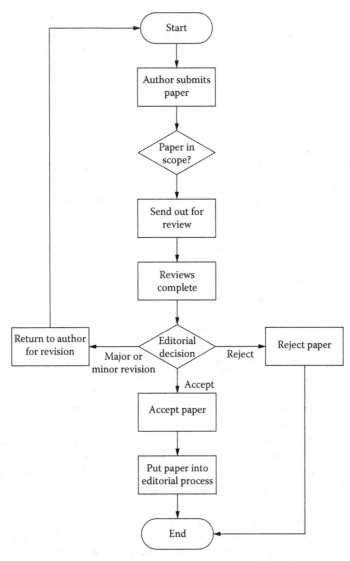

FIGURE 8.1
Simplified paper review process.

8.3.3 Review of Books

The review process for books is slightly different. A book proposal typically consists of an outline of the proposed book, a competitive analysis and market survey, a detailed author profile, and one or two sample chapters. A potential author prepares a book proposal and it is then reviewed by independent reviewers and the publisher's editorial board. The proposal is then accepted or rejected based on these reviews and the publisher's own market research.

8.4 Handling Rejection

If you are going to try to publish work, in any venue, be prepared for rejection. Every writer, no matter how famous, learned, popular, etc., will experience rejection. Stephen King's *Carrie* was rejected 30 times, *Gone with the Wind* by Margaret Mitchell was rejected 38 times, and even J.K. Rowling's first *Harry Potter* book was rejected 9 times [Maddox 2008]. Of course, these were not technical writers, but their persistence is to be admired and emulated. With experience, your work will get rejected less often, and you will become better at dealing with rejection, but it always hurts to have your work returned with harsh criticism.

8.4.1 Rejection Letters

It is always disappointing to have your work rejected, even when the reasons given are fair. For research papers, rejection letters comprise a summary of the reviews and provide useful information for improving the manuscript. For example, the reviewers may indicate some missing research in the literature review, or they may find a flaw in the research methodology. Sometimes the reviewers simply want you to reorganize your work so that it is more easily understood. Whatever the case, and no matter how coarsely worded the criticisms may be, you must try to find the good in the comments from the editor and reviewers.

But sometimes the criticism makes no sense, as if the reviewer hadn't read the paper. Often the rejections are unfair—it might be clear that your work wasn't fairly reviewed and that you received a *pro forma* rejection. Rarely, the rejection letters can be inappropriately personal even including name-calling, and there is no place for that.

So, you must be psychologically prepared to have your work rejected for both fair and unfair reasons. You just need to know that the scientific and technical publication business is very rough, and it can be hard to get your work accepted. There is an irony, too, in that very bad work can get published while good work does not.

You should always take the legitimate feedback from a rejection, improve your writing, and resubmit to another venue, but never let rejection make you stop trying to publish. I have had many papers and books rejected multiple times, but each time, I would make improvements and resubmit the work until it was published (OK, in a few cases, I finally gave up). *Easy PC Maintenance and Repair* [Laplante 1992] was rejected 20 times. Some of the rejections were *pro forma* (see Figure 8.2), while others were encouraging (see Figure 8.3). But the book was finally bought by a publisher.

8.4.2 Responding to Rejection Letters

Here is some advice on dealing with rejection letters: If the paper was rejected, politely thank the EIC and think of new ways to improve the work and other venues in which to publish it. Don't write back to the editor in an angry or defensive tone. You don't want to anger the editor because most scientific communities are relatively small and malcontents will become pariahs. Moreover, being nice can endear compassion from the editor and perhaps help with your next submission.

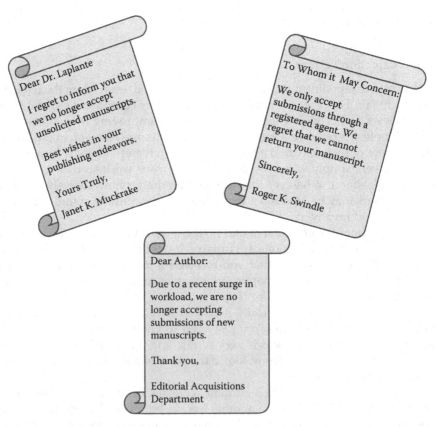

FIGURE 8.2
Stylized sample rejection letters for my first book proposal.

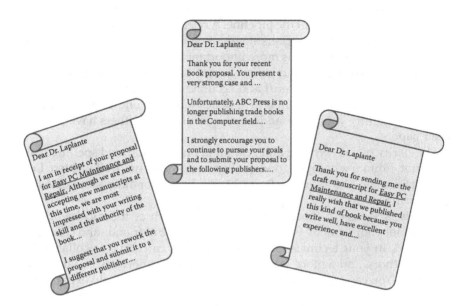

FIGURE 8.3
Stylized examples of more encouraging rejection letters I received for my first book proposal.

Ask for clarification and guidance, if the paper has been judged to be worthy after revision—persistence shows the editor that you are firm in your convictions. But don't push too hard, even if you are convinced that you are right on some point and the reviewers are wrong. Compromise if you must and look for win–win situations. For example, if a paper is rejected because it is too long, perhaps it can be broken into two shorter papers.

Try to learn from your mistakes. If your papers are getting rejected time and again because of poor writing, an incomplete analysis of previous work, or not enough data, then it is your own fault. Still, not every paper will be acceptable, even after being revised several times; so learn when to cut your losses. My effective success rate is about 80% for papers, meaning that for every eight papers I published, two others simply could not be made acceptable for any publishing venue and had to be discarded.

8.4.3 Succeeding at Publishing

Finally, here is some general advice that will help you increase your publishing success rate.

Do favors for the journals and publishers; for example, volunteer to be a reviewer. Journals and magazines are always looking for eager, qualified reviewers. You will learn more about the publication as a reviewer than you do as a reader or author, especially about what kind of material and level of quality are acceptable for the publication.

Keep multiple projects in the pipeline. A rejection of one paper stings less if you have other papers in various stages of review, revision, and publication.

The quality of your work is more important than the quantity. In addition to technical correctness, your writing must have a unique contribution or point of view. The most careful and accurate research results are going to have difficulty finding a venue for publication if no one finds them interesting. Hopefully your work addresses questions that others have posed in their papers.

The clarity of your writing and grammar, punctuation, and spelling really do count. Editors will reject good quality work that is not well written. Look back at Sections 3.1.3 "Revising" and 3.1.4 "Editing" for more about polishing your written work.

Finally, it is very difficult to do quality research or technical work and write about your work by yourself. Build a network of friends and colleagues who work in your technical area. These friends can serve as collaborators and coauthors, and a more seasoned member of the network could become a mentor. You can tap into the network of these friends to help expand your visibility and knowledge of opportunities for special issues of journals and magazines in your area of expertise.

8.4.4 Vignette: Experiences with Reviews

Over many years of research and writing I have had hundreds of papers peer reviewed. In most of these reviews, even when my papers were rejected, I have received helpful criticism. Sometimes it was evident that the reviewers had not carefully read the paper, for example, they would ask for inclusion of an idea that was already included. But, on a few occasions, I have received unnecessarily harsh, even bizarre feedback.

For example, on one occasion a reviewer recommended rejection of my paper simply because "He is not really a faculty member at Penn State." This assertion is incorrect, of course, but my home location is not at the main campus—a fact which the reviewer perceived as some form of deception. On another occasion, my paper was rejected from a journal for being out of scope (meaning, not with in the focus area) of the journal by a guest editor. The problem was, however, that the journal was one in which I was the founding editor—I had defined the scope of the journal. Finally, I had a paper accepted to a journal and then during the production process, a new EIC was appointed who promptly rejected the previously accepted paper.

The point of recounting these stories is not for sympathy—as I have suggested everyone who writes papers that are reviewed will encounter rejection. The lesson from these experiences is to reinforce the idea that if you are going to try to get work published, you have to be ready to deal with all kinds of rebukes, even unfair ones.

8.5 Open Access Publishing

8.5.1 The Traditional Publishing Model

When publishing your work for commercial purposes, whether it is a book or an article in a journal, there is an intellectual property right called "the copyright," which the creator of the work owns unless it is transferred (given or sold) to another person or legal entity in exchange for some benefit. For example, I wrote this book and then sold the copyright to all the material herein (for which I owned the copyrights) to Taylor & Francis in exchange for editing, formatting, binding, advertising, and distributing this book and a "royalty"—a percentage of the net sales revenue of the book.

In the case of works in scholarly journals and magazines, my benefit in selling the copyright is the editorial services and wide distribution of my ideas, the latter being very important to a professor. It costs a lot of money to process scholarly manuscripts. Manuscripts must be reviewed by peer experts. The reviews must be managed and considered in order to render an editorial decision. If the manuscript is accepted by the reviewers, then there is a great deal of editing, formatting, redrawing of figures, and so forth that must be done. The manuscript needs to be tagged with metadata that allows it to be archived and discovered by search tools. And all manuscripts must be organized into a journal or magazine in a logical way for each issue.

While some of the editorial work is done by volunteers, there still must be paid professionals to oversee the process. Therefore, most publishers sell these scholarly journals and magazines to libraries, corporations, and individuals in order to support publishing activities and to make a profit for shareholders.

Even in the case of the publishing operations of nonprofit organizations, such as the Institute of Electrical and Electronics Engineers (IEEE), which publishes more than 150 journals and magazines, the operating income from these periodicals must be net positive in order to fund the other worthy activities of this nonprofit entity. As a reader, author, or reviewer, it is not always apparent whether a publisher is for-profit or nonprofit.

8.5.2 The Open-Access Publishing Model

Scholarly books, journals, and magazines are expensive, and there is a growing movement that declares that "information should be accessible for free," particularly when this information was supported in part or in full by grants from government entities. This worldwide movement believes that scholarly publications should be made available to the public via "open access," or through organizations that do not charge for access to the information.

A more precise statement of open access was formulated at a 2003 meeting of open access proponents in Bethesda, Maryland, and is therefore referred to as the "Bethesda Statement" [Bethesda 2003]. The Bethesda Statement requires that in order to qualify as open access, the intellectual property holders must grant the right for anyone to use the work in electronic form in any way, and to make "small numbers of printed copies for their personal use." The other requirement is that the work be placed in an unrestricted digital library that meets certain conditions of support. The Bethesda Statement does not give permission to make large-scale reproduction and distribution of the work in print form, recognizing that it is costly to make and distribute print content. Open access publishing does not mean free publishing, although anyone can give away information for free, for example by simply placing the work on a public website. For high-quality work, there is a cost to producing, editing, drawing, and typesetting that information—costs that someone must bear. People are willing to pay to subscribe to scientific and technical journals and magazines because the information has value. Economic theory asserts that giving away information for free implies that the information has no value, or that there is an ulterior motive.

The costs of managing the editorial review process, formatting, tagging, etc. do not change for open access publications. These costs must be borne by some entity. If they are not being paid by users of the information, that is, the readers, then they are paid by the producers of the information, that is, the authors or their employers and sponsors, or by the taxpayer. This financial model is very different from the traditional one, and it presents several challenges to for-profit and nonprofit publishers, mostly with respect to budgeting.

There are two models for open access publishing: "green" and "gold." In the green model, authors post their work through a publisher's freely accessible database. Some traditional publishers allow authors to post their work in a public repository after an embargo period has expired (at which time it is assumed the publisher has recouped its costs). In the gold model, publications are made open access immediately through publishers' websites with the authors paying an article processing fee. The fee can be substantial, usually in the range of $1000–2000.

In the last few years, certain for-profit, open access only publishers have emerged, which follow the gold model (see Section 8.5.5). Many traditional publishers also offer a hybrid open access option by allowing the authors (or their sponsors) to pay the processing fee in order to make the paper publically available. Why would authors pay to publish their own work? One reason is that they have been funded by a university or government agency to produce the work, and one of the conditions of that arrangement is that they make their information freely available. Other motivations for giving away high-quality work for free include wide exposure of the ideas, enhanced reputation for the authors, or for humanitarian purposes.

I believe that the open access publishing model cannot yet be declared a success. For example, it is not clear that academic institutions value open access publications as much as traditional subscription-based access model, particularly with respect to tenure and promotion decisions. It is also unclear if the financial model for open access is viable (see Section 8.5.5).

8.5.3 Vignette: Experience with Open Access Publishing

I was a member of the editorial board of an open access publication, *Advances in Software Engineering*, published by Hindawi Publishing Corporation (http://www.hindawi.com/journals/ase/). Hindawi is a for-profit company based in Cairo, Egypt, and New York City and they publish good quality journals. As an editor, I handled paper reviews and organized a special issue. The quality of the work that I approved was quite good, and from this experience, I can say that the rigor of the process and the quality of papers in these open access journals can be as good as in traditional journals. The results depend on the diligence of the editor(s) managing the reviews and making editorial decision.

Unfortunately, *Advances in Software Engineering* ceased publication in 2017, presumably, due to lack of submissions. The published papers are still available in the archives of Hindawi.

8.6 Self-Publishing

There are several mechanisms for self-publication. The two most common self-publication mechanisms are vanity presses and online self-publishing. A third mechanism is to form your own publishing company. I have seen the latter done by friends a few times, but their negative experiences convinced me that this approach is not viable.

8.6.1 Vanity Presses

If you are desperate to publish your book in a quasi-mainstream form, then you may opt for self-publication using a "vanity press." A vanity press is a commercial publishing firm that underwrites its publications costs by charging the author for the right to publish, rather than investing up-front in the production costs, while hoping for payback after publication. Because the author subsidizes the publication costs and bears the financial risk, the review process is much simpler and the threshold of quality and significance is that much lower. In other words, a vanity press is anxious to have you as its customer.

Depending on how much the author pays, the vanity press will perform one or more of the following services for the author on an á la carte basis:

- Copyediting
- Layout
- Re-rendering art
- Indexing

- Binding and printing
- Limited marketing (usually by including a listing for the book in the publisher's print and online catalog)

The vanity presses pay higher royalty rates and can sell at lower prices than traditional publishers because they have less of an investment to recover. But the potential for selling books in quantity is greatly diminished because the vanity publisher conducts very little marketing.

In vanity publishing, most of the risk is placed on you, the author. In addition to the time you invested in writing that book, you will have to pay a vanity press on the order of $5000 to $10,000 before your book will appear in print. Even at a high royalty rate, you will have to sell many books to recover those costs and make any kind of profit.

If all you care about is having a book in your hand with your name on it as the author, and you can't get your book published through traditional publishers, then go ahead and publish with a vanity press publisher. Edgar Allan Poe, Stephen King, and Mark Twain were self-published before becoming famous [Ormrod 2016], perhaps you will be too.

8.6.2 Online Publishing

The simplest form of self-publishing is to write the book, do the layout with some commercial software, and then make the book available through your own website. While this is an easy way to self-publish, the big problem is marketing: Who is going to find your book, and when they do find it, why should they purchase it? *Refactoring to Patterns* was self-published this way, and it was such a good book that it eventually was acquired by a mainstream publisher [Kerievsky 2004]. This book is selling very well too.

There are Web-based services that will help you publish your book through a variety of á la carte services, such as editing, layout, graphical design, indexing, and so forth. They can deliver your book to purchasers electronically or as a bound printed copy. These publishers also provide a more sophisticated Web presence and some marketing for your book—but at the cost of sharing the royalties. However, I think these costs are well worth it to the novice author. I would rather publish an online book through one of these companies than try to do it myself. Self-publishing online is a good alternative to vanity publishing if you cannot get your work published by a mainstream publisher.

You can also publish your book in print and for Kindle via Amazon.com and other e-commerce sites. Jacobs [2014] provides a good discussion of these possibilities and Chapter 9 further discusses electronic publishing.

8.6.3 Vignette: Bootleg Books

One of the biggest threats to the profits of book publishers and authors is illegal copies being sold (or given away) outside the region where they are permitted to be sold. Many publishers will produce low-cost copies of books for sale outside the United States. The costs are kept down by using lower-cost papers and inexpensive soft covers.[4] The rights to print these low-cost books are then sold to a distributor at a relatively low one-time cost with a restriction on the book's distribution limiting the distributor's sales to just one region. The book will carry a warning, for example, "Not to be sold outside of India and Pakistan." The author receives a percentage of this one-time fee, usually 50%. But the fee is very low, often $1000 or less.

For example, I received a $500 royalty for the sale of the Asian rights to one of my books. The publisher that bought those rights can print tens of thousands of copies or more. However, these low-cost copies are often resold, illegally, via the Internet outside the permissible region of distribution.

I personally witnessed this black market activity. I was teaching a course using the aforementioned book, and I noticed that some students had "Asia-only" copies. The students told me that they purchased the books via the Internet. The students didn't know that the books were intended for sale in Asia only, so I didn't blame them. But the publisher must account for this kind of illegal redistribution, and sets its U.S. prices accordingly.

Now you might think that a publisher should charge more for the Asian distribution rights or not sell these rights at all. But if the foreign distribution rights are priced too high, then the Asian publisher will not buy the rights, in which case bootleg copies of my book will appear in Asia. If the publisher tries to sell the book in its high production value form, customers in Asia will not buy the book, and again, bootleg copies of the book will begin to appear. So, the publishers have determined that the best strategy is to offer low-cost alternatives in certain markets and hope that black market sales of the book are minimal.

If you are going to self-publish, be prepared, as you might find your book bootlegged. And please, don't purchase bootleg books.

8.7 Exercises

8.1 Locate online job advertisements for "technical writer" or some variation of the term. What do the job qualifications for each advertisement have in common?

8.2 Search the Internet for freelance writing jobs. What kinds of jobs and projects do you see? What kinds of rates and fees do these projects pay?

8.3 Search the Internet for open access journals and describe the business model of the publishers (i.e., for-profit versus nonprofit).

8.4 Does your favorite academic library include open access journals in its catalog of periodicals? Why or why not?

8.5 Search the Web for vanity publishers and self-publishing sites. Describe your findings. Which of these publishers seems the most likely fit for your purposes?

8.6 Write a two-page essay on the U.S. National Institute of Health's "Public Access Policy." In particular, address its origin and what it means to researchers and publishers.

8.7 Find a book technical book that was originally self-published but eventually was published by a mainstream publisher. Briefly describe the situation.

8.8 "Open source" refers to software or documentation that is made available for free use and redistribution provided that certain license provisions are followed. For open source artifacts, the licenses that permit free use and redistribution of intellectual property contain "copyleft" provisions. Write a two-page paper discussing the many forms of copyleft rights.

8.9 There are some who argue that creating and distributing bootleg books is necessary, since technical books are expensive and the high prices discriminate against researchers and students in poorer countries. Write a two-page essay taking either the side in favor of bootlegging or against.

8.10 How do you deal with failure or rejection (personal or professional)? Write a two-page essay on the topic giving examples and creating summary.

Endnotes

1. I have written four introductory books [Laplante 1992, 1993, 1995; Laplante and Martin, 1993]. I find these the most satisfying to write, but they have been the least profitable for me.

2. A college bookstore charges more than another retailer because the bookstore has to pay a fee to the host college. The college bookstore likely has a more generous return policy than a retailer in the case of cancelled courses or too many copies. And the college bookstore is much more convenient, especially if you wait to hear the professor's first lecture to decide if you are going to stay in the course and buy

the book. The college bookstore has to account for these factors in its pricing.

3. A version of this story first appeared in Laplante [2008].

4. You may ask, "Why not sell these low-cost versions in the United States?" Publishers have found that U.S. buyers don't like books produced with low-cost production practices and will avoid buying them if an alternative book at high production values is available.

References

Bethesda Statement, Summary of the April 11, 2003, Meeting on Open Access Publishing, http://legacy.earlham.edu/~peters/fos/bethesda.htm, accessed December 31, 2017.

Gilreath, W. and Laplante, P. A., *Computer Architecture: A Minimalist Approach,* Kluwer Academic Press, Boston, MA, 2003.

Jacobs, D., How to Self-Publish Your Book through Amazon, *Forbes.com,* http://www .forbes.com/sites/deborahljacobs/2014/04/25/how-toself-publish-your-book -through-amazon. April, 2014, accessed December 12, 2017.

Kerievsky, J., *Refactoring to Patterns,* Addison-Wesley, Boston, MA, 2004.

Laplante, P. A., *Easy PC Maintenance and Repair,* Windcrest/McGraw-Hill, Blue Ridge Summit, PA, 1992.

Laplante, P. A., *Fractal Mania,* Windcrest/McGraw-Hill, Blue Ridge Summit, PA, 1993.

Laplante, P. A., A reading list of classic papers for computer science majors, *Mathematics and Computer Science Education,* 28(20), 198–204, 1994.

Laplante, P. A., *Easy PC Maintenance and Repair, Second Edition,* Windcrest/McGraw-Hill, Blue Ridge Summit, PA, 1995.

Laplante, P. A., (Ed.), *Great Papers in Computer Science,* West Publishing, St. Paul, MN, 1996.

Laplante, P. A., *Real-Time Systems Design and Analysis, Third Edition,* John Wiley & Sons/IEEE Press, Hoboken, NJ, 2004.

Laplante, P. A., Great papers in computer science: A retrospective, *Journal of Scientific and Practical Computing,* 2(10), 31–35, 2008.

Laplante, P. A. and Martin, R., *Using Unix,* West Publishing, St. Paul, 1993.

Maddox, M., Famous Books Rejected Multiple Times, *DailyWritingTips,* https://www .dailywritingtips.com/famous-books-rejected-multiple-times/, December 2008, accessed January 1, 2018.

Ormrod, J., 6 Famous Authors Who Chose to Self-Publish, *IndieReader,* https:// indiereader.com/2016/10/6-famous-authors-chose-self-publish/, October 12, 2016, accessed January 1, 2018.

9

Writing for E-Media

9.1 Introduction

As the costs of printing and maintaining inventory increase, content producers are looking for new models to create and distribute technical material. I consider "technical materials" to be books, technical papers, marketing and advertising content for technical products, user manuals, newsletters, safety reports, and so on. The producers of these materials include traditional publishing houses, companies, government agencies, and every other kind of entity. To lower costs, the producers of this written technical content rely increasingly on direct publication via the Web or through electronic media. In addition, because of the efficiency of electronically storing information and the advantages of archiving and searching, there is strong motivation to rely more heavily on distribution via e-media. E-media includes CD-ROMs, small portable memory storage devices (e.g., memory sticks), electronic readers (e-readers), and Web-based hosting of content.

Web-based publication venues may bypass print publication entirely. These mechanisms include electronic-only magazines (e-zines), corporate websites, social networking groups devoted to some technical topic, pundit blogs, and Twitter accounts. Writing for e-media is a unique form of writing that has certain advantages and some disadvantages. The advantages include:

Lower overall cost

Potential wider exposure to ideas

Ease of distribution to rural and remote areas

Ease in correcting errors discovered post publication

Capability to publish much more quickly than with print media

Capability to link related content and direct readers to other material of interest

Discoverability (on the Web by crawlers) and classification (if appropriate tagging is done)

Lower cost to include color content and high-fidelity graphics

Capability to embed dynamic content

Environmentally friendly

Some researchers even contend that accessing material via e-readers "increases the pleasure found with its use and the extent to which it is actually used" [Antón et al. 2017] though their study did not focus on technical writing.

There are disadvantages to publishing content on e-media too. These include:

Increased vulnerability to theft of intellectual property (e.g., plagiarism, unauthorized copying, and distribution of published material)

Unwanted widescale exposure of proprietary or secret information via hacking

Widescale exposure to potentially embarrassing mistakes

Low barrier to entry for competition (i.e., anyone can publish to a website)

One study found that "people highly involved with reading tend to perceive e-book readers as useless, which hampers their adoption" [Antón et al. 2013]. But despite these potentially game-stopping disadvantages, use of the Web has exploded due to the dramatic increase in content coming from publishers, companies, and individuals.

In Chapter 9, I want to focus on writing for electronic publications as a special form of writing and also share my experiences in writing for various forms of e-media, including e-mail.

9.2 E-Mail Can Be Dangerous

An electronic mail (e-mail) message is a very desirable form of business and technical communication. E-mails can have the same informality as an in-person or telephone conversation, or the formality of a legal contract. Unlike a conversation or phone call, however, e-mails are persistent; they can be saved and produced as evidence that you said something that you claim you did not say. Therefore, you must write e-mails carefully, and it is a very good idea to "incubate" them before sending them in haste.

Do not use e-mails as a substitute for in-person communications, however. Firing or breaking a relationship with someone via e-mail (unless time and distance necessitate) will be perceived as somewhat cowardly. If a written confirmation of some interaction is required, but in-person communication is preferred, I suggest you use both. That is, have the face-to-face conversation and then follow up with a written confirmation of the meeting, either via e-mail or in a formal paper document.

Some e-mail software does not support equations, and therefore e-mail is not always the preferred mechanism for technical communications.

9.2.1 Rules for E-mails

I only want to address the use of e-mail as a form of communication for business and technical matters. The various issues in using e-mail as a personal form of communication are beyond our scope here. I won't give you a template for an e-mail as it is an informal communication medium and must be adapted to the situation and task. But for technical e-mail, it is worth adhering to some basic principles.

In a professional setting, e-mails should be short, concise communications containing some call to action or acting as a cover letter for a longer, more complex attached document. Here are some other general rules for writing effective e-mails:

1. Include relevant keywords in the subject line (for ease of searching and organizing stored e-mails).
2. Keep the e-mail brief. The entire e-mail should fit inside the space of one screen.
3. If appropriate, the e-mail should contain a call to action (e.g., "Please let me know of your intentions by 9/12/2018, 9 am EDT").
4. Use "receipt confirmation" features sparingly—it annoys the recipient to have to click "confirm" upon opening some trivial e-mail.
5. Use "urgent" tagging of messages sparingly. Truly urgent messages are rare.
6. Use simple writing: avoid flowery language and needless complexity.

The above suggestions should be customized to conform to the local policies of your employer, school, or organization, particularly with respect to civility and propriety.

9.2.2 The Signature Line

The format of your signature line can convey subtle information about you. Too much information and embellishments might brand you a snob. But too little information is not good either.

I like to vary my signature line, depending on the situation. For example, my full academic e-mail signature line looks like this:

Phillip A. Laplante, CSDP, PE, PhD
Professor of Software and Systems Engineering
Affiliate Professor of Information Science and Technology
Fellow IEEE, Fellow SPIE

Penn State
www.personal.psu.edu/pal11

I use the above form of signature in most of my e-mail transactions because this is a conventional format for a professor and it contains important information. But I could compress this signature line significantly. Remember John Thompson from Chapter 2? His verbose sign was distilled to a very simple one with a hat icon and his name. I can use a similar kind of editing compression for my signature line, and I could use the various forms depending on the situation.

For example, I can remove my courtesy appointment in another college of Penn State ("Affiliate Professor of Information Science and Technology") and remove my honorific Fellowships from two academic societies. These embellishments are highly regarded in academia, but might be perceived as hubris in industry. Anyway, you can find these listed on my website if you follow the link provided. Now my signature line looks like:

Phillip A. Laplante
Professor of Software and Systems Engineering

Penn State
www.personal.psu.edu/pal11

If I am writing some personal e-mail, then why list my job title and employer? Then my e-mail signature would look like this.

Phil Laplante
www.personal.psu.edu/pal11

If I really want to close the e-mail informally, I could use the following:

Phil
www.personal.psu.edu/pal11

My family, friends, and close colleagues don't need to be constantly reminded of who I am, so I can omit the Web link, leaving the most informal closing:

Phil

When I communicate with my students I use the following signature:

PAL

which are my initials. I use this form because I discourage students from addressing me by my first name. This practice is not egotism—I discovered long ago that students place a higher value on professors who are addressed formally than those who permit informal interaction.

There are even times when a signature is unnecessary; after all, my identity is contained in the header of the e-mail. Whatever signature line you use, choose it carefully and customize the signature for the situation in which it is used.

9.2.3 Use of Emoticons

Emoticons are text-based graphics that depict facial expressions in order to represent an underlying emotion. Emoticons have apparently been used since the mid-to-late 1800s in print writing [Lee 2009]. I recall seeing and using emoticons on teletype printouts and on paper punch tape outputs in the 1970s.

Individuals seem to use emoticons to simulate "face-to-face communication with respect to social context and interaction partner" [Derks et al. 2008]. Emoticons can help convey emotion, strengthen a message, and express humor over the sometimes stark communication medium of e-mail.

You can find many clever and complex emoticon sets all over the Web. For example, there is a large and ever-changing set of emoticons on the Wikipedia site [Wikipedia 2018]. I tend to use a very simple subset of these emoticons, as shown in Table 9.1.

One of the more common uses of emoticons is to overcome the tendency of e-mail to be perceived as formal or serious when the writer, whose facility with words may be less than superb, intended humor, or even sarcasm. If you find that you must resort to the use of emoticons to soften e-mails, then this should be a warning signal.

TABLE 9.1

List of Emoticons I Like to Use

Emoticon	Facial Expression Represented	Emotion/Reaction Represented
:)	Smile	Happy
:(	Frown	Sad
;)	Wink	"Just kidding" or "Isn't that cute"
:o	Agape	Shocked or surprised
;(	Scowl	Angry

For informal professional e-mail and personal communications, emoticons are great, but I don't advise using them in official business communications. If there is a sentiment that you need to convey, it is best to convey that sentiment in person, or if distance represents a barrier, then via telephone as a last resort. I think more complex emotions should be communicated in person or via a telephone call.

Once you send an e-mail with a strong sentiment, you can neither take it back nor deny its contents. At least with a personal conversation or phone call, you can explain or otherwise downplay your emotional state later. Also, facial expressions and tones of voice can provide additional depth to your message, adding clarifying complexity and subtlety that is missing from e-mail.

9.3 E-Newsletters

A "community of interest" is some group of like-minded individuals who chose to associate via in-person meetings, mail, or electronic means. A community of interest can be established for any type of shared professional, recreational, political, religious, or personal interests. For example, there could be communities for "Beet Lovers," "Ugly Dog Fans," "Klingon Speakers," and "Left-Handed Guitarists." Some of these unusual communities do exist (see "Exercises").

Technical communities of interest are usually centered on a particular technology, product, or company. Whatever the community, the Internet provides a great way for a group to connect over time and space via social networking sites, blogs, e-mail, or e-newsletters. An e-newsletter is a simple periodical that is distributed exclusively via an e-mail server or may be downloaded from a webpage.

An electronic or e-newsletter can have any format and can include elaborate graphics, although including graphics may raise compatibility problems with certain e-mail client programs.

A good e-newsletter should be short and packed with information and calls to action. If the e-newsletter pertains to a group that meets physically, then there should be ample information to allow members to find the next meeting.

E-newsletters are quite informal and generally lack fancy formatting (see the vignette on the CIO Institute, for example). The main features are announcements, contact points, event placeholders, and "save-the-date" type items. The main goal of the newsletter is to keep the rhythm of the community going and to disseminate regular information regarding the activities of the group. Community members can always get more information from each other, from the community's website, or directly from the editor of the e-newsletter. For your convenience, an e-newsletter template is given in Appendix B.

9.3.1 Vignette: A Newsletter for CIO Institute

For almost ten years I wrote and distributed a monthly e-newsletter to a community of practice for chief information officers (CIOs). Let's call that community the "CIO Friends." Here is a disguised version of one of the CIO Friends newsletters from a few years ago.

CIO FRIENDS NEWSLETTER #46

FEBRUARY 2015

WELCOME NEW MEMBERS

Arthur Kinnealy: Bako Pharmaceuticals
Susan Graham: Anderson Health System
Cheryl Jones: Billingsly Software
Red Scanlon: MuffinMakers

ANNOUNCEMENTS

Amy Chin reports that the Convention Hall will be offering free high-speed wireless access to the Internet in all public areas. We will be holding a future event there so you will be able to stay connected when you attend.
Please update your calendars (see attached program doc) as some of the event dates have been changed.

EVENTS

For more information on any meeting, please either visit the CIO Friends website (www.ciofriend.org), or contact Kerry Winston at kwinston@ciofriend.org. To attend, please RSVP to Kerry Winston.

APRIL

Wed., April 21: "The Data Center Dilemma: Buy vs. Build" (Uniflow Corporate Headquarters, 81 Market Street, Philadelphia; 8 a.m. to 10 a.m.). This event, hosted by Uniflow, features Uniflow cofounder and CEO, Roger Pester. At this roundtable event you will learn how to calculate data center build costs; the importance of creating a world-class $24 \times 7 \times 365$ support system; and where to look for hidden data center management costs. See the attached program document (datacenter.doc) for more details.

Wed., April 28th: "New Technologies, New Solutions Rethinking Business Intelligence for 2021 and Beyond." Features Sally Garfinkle, VP of Alliance Consulting. Sally is a leading authority on Data Warehousing and Business Intelligence and has written five books on Data Warehousing. Many of you may recall Sally's well-received talk on Data Warehousing last year. The event, which is sponsored by RoeTastic Consulting, will be held at the Raynor Hotel from 8:30 a.m. to 10:30 a.m. and will include a continental breakfast. See the attached program for details (CIOFriends.pdf).

MAY

May 11th Data Migration: "A Success Story in Reality," sponsored by Decadent Software. Our speaker/panelists will be Fred Reed, Executive Director of the Advanced Solutions Group at Decadent and Winifred Hochule, CTO of General Laboratories. This event is a follow-up to the April meeting with Sally Garfinkle. The meeting will be held at the Rayburn Hotel, 8:00 a.m. to 10:30 a.m. More information will be forthcoming.

May 19th at Bankbook Trust: This roundtable discussion focuses on the Changing Role of the CIO. Ganesh Mitra, a former President, COO, and CIO, and currently a venture capitalist, will lead the discussion. See the attached document (changing.doc) for more details.

JUNE

June 3 "IT Procurement and Outsourcing Do's and Don'ts: A Trial Lawyer's Perspective": A by-invitation-only CIO Friends event intended for top IT decision makers, CIOs, and CEOs. Sponsored by Dewey, Cheatham, and Howe (Penn State Great Valley, Malvern; 8:30 a.m. to 10:30 am). Our speaker is Les Dewey, cofounder and managing partner of Dewey, Cheatham, and Howe and a specialist in information technology litigation. See attached program (philly.doc).

JULY

July 14, location TBD: This roundtable discussion focuses on IT architecture. A guest moderator (to be named) will lead the discussion.

MISCELLANEOUS

Please keep sending me your ideas for speakers, programs, events, and sponsors. Also, please keep visiting the CIO Friends website (www.ciofriends.org) and CIO Friends link (www.ciofriends.org/cio.cfm) for more information on upcoming events.

This example illustrates the key elements of an e-newsletter. A template for an e-newsletter can be found in Appendix B.

9.4 Blogging

The term "blog" is a mash-up of the words "Web" and "log" (say those words fast and you hear "web-blog"). Blogs feature short, timely, and informal information snippets. Technical blogs abound.

Blog entries should be timely, informal, and informative. Even in technical blogs, the tone should be conversational so that you can attract and retain a regular audience of readers. For example, a blog is a good way to handle "frequently asked questions" for a technical product. A group of users of some technology can share their experiences, tips, tricks, and so forth via blogging.

You can start your own blog easily, for example on Wordpress (www.word press.com) or Tumblr (www.tumblr.com). These sites provide platforms for bloggers, often for free (but supported by banner advertising).

I used to share a blog that was hosted by a major IT publication, *IT World*. My co-blogger[1] and I would try to alternate and post interesting information on a regular basis. The following is one of the blog entries that I wrote:[2]

Driving Business Value Through IT During Tough Times

October 8, 2018, 2:30 pm

Yesterday, Phil participated in a CIO roundtable discussion on "driving business value through IT." While the roundtable was planned months in advance, no one could have anticipated how the discussion would be dominated by the latest economic woes. Most of the CIOs were feeling tremendous cost cutting pressure—beyond the usual levels, and several had given out pink slips that morning.

Some of the main themes that emerged were—maintaining a trust level by delivering on small promises early; showing the true ROI of IT initiatives, whether they be new or on the chopping block; and keeping staff motivated (and retaining them) when IT cost-cutting initiatives turn the office of the CIO into the office of "boredom."

Phil's recommendation: don't thwart innovation through constant focus on cost cutting and ROI calculations. Maintain a sandbox where innovation can be safely tried—and give employees time to experiment with new tools and ideas. Innovation needs to be encouraged, not relegated to the employees' own time at home. That's how you drive business value through IT innovation—even in tough times [Laplante and Costello 2008].

Blogs are notable because they are contemporaneous, but as a result, the information has a limited useful life span. My coblogger and I found that although we only needed to write short pieces, it became difficult to do so and we just couldn't post information fast enough to create a regular audience for our blog.

If you are interested in starting your own blog, be careful of legal restrictions where you live. For example, at one time the city of Philadelphia required resident bloggers to pay a $300 business license [Hemingway 2010].

9.5 Social Networks

Social networks are communities of interests enabled by some unifying Internet technology. Social networking sites such as LinkedIn and Facebook are growing in both personal and professional usage. With more than 2.2 billion members, Facebook is the largest social networking service (see Figure 9.1). Widely used social networking sites such as Facebook and LinkedIn provide exchange forums for technical ideas and business and social networking for technology professionals. Many technology enterprises are incorporating social networking sites as part of their marketing, sales, post-sales support, and recruiting strategies. These companies use Facebook to post pictures or video of its products and customer feedback and suggestions, Twitter updates are used to announce sales, and LinkedIn helps connect key business partners. Social networking capabilities have even been extended to mobile network devices, and there are solutions that support photo sharing and cross-network mobility [Laplante 2010].

Many professional societies, for example, the IEEE are using virtual communities to grow and sustain membership. Although face-to-face contact

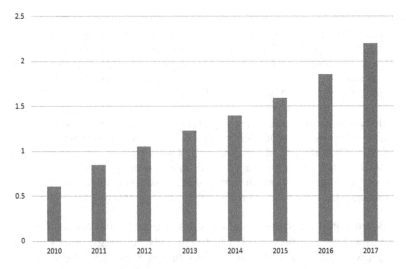

FIGURE 9.1
Billions of active Facebook users (data from http://internetstatstoday.com/).

is still valuable, younger professionals have become used to interacting virtually and the costs to create communities of interest is much lower in the virtual world. Professional societies and corporations have taken note of this trend, and are adapting [Milojicic and Laplante 2011].

9.6 E-Magazines[3]

Electronic or digital magazines (e-magazines or e-zines) are the equivalent of a printed, glossy technical trade magazine or refereed magazine, only in electronic format. E-zines can be read on a laptop, desktop, smartphones, or other personal digital devices. E-magazines seek to provide an enhanced reading experience that is so compelling that readers want to return to it frequently and share it with others.

In principle, e-zines have an attractive list of features, including:

Content searching

Ease in making annotations

Linking through the table of contents

Computer-readable formats

Linking to references

Easy forwarding

Frequent updating

Bookmarking

Page notetaking

Other features include

Access to Web articles

Links to discussion boards

Author blogs

Videos

Animations

Social networking linkages

Audio

Simulated page turning

These devices are also environmentally friendly in that no paper or ink is used.

There are three general types of e-magazines: pdf, medium enhanced, and fully enhanced. The simplest is the pdf magazine, a visual equivalent of the print version of the same magazine. The pdf version, however, can have a linked table of contents, be searchable, and include links to external content. The electronic version of the IEEE Computer Society's *Computer* magazine is one such example.

A "medium-enhanced" e-zine has all the features of the pdf e-magazine plus additional written content (not found in the print version), social media support, and extra links. The electronic version of *PC Magazine* is an example of a medium-enhanced e-zine.

Finally, "fully enhanced" e-zines have all the features of a medium-enhanced e-zine, including support for video, audio, animation, feedback, and newsfeeds. Technical magazines such as *Popular Mechanics* and *Wired* have fully enhanced e-zines.

Successful digital magazines provide convenience, currency, community, multimedia content, and the ability to provide additional content beyond the constraints of a limited number of printed pages. Technical writers should consider that their words are likely to find their way into this digital form.

9.7 E-Readers[4]

An electronic reader (or e-reader) is an electronic device designed for reading digital books (or e-books) and periodicals available in electronics form. E-readers use special electronic typesetting technology, called e-ink, to display content to readers. E-readers are portable, can be read even in bright sunlight, and have a relatively long battery life. Some personal digital assistants (PDAs) have e-book reader–like properties.

9.7.1 Common Features

Common features of e-readers include:

Inputs through buttons, stylus, or touch

Loading of e-books via USB connection to computer or direct Internet access via Wi-Fi or 3G

Capability of annotation and notetaking

Conversion of content into text via third-party software on using another computer

While the number of e-reader devices increases rapidly, there are some differences in the availability of devices and their features, depending on your location. For example, important features (such as folder management, tagging, search, and bookmarking) are not always available on all devices. Display sizes vary too—most are six, eight, or ten inches, although larger screens are expected.

E-readers can support a wide variety of "readable" file formats, including the standard formats (such as txt, rtf, doc, and pdf) and specialized e-book formats (such as Kindle mobi, Adobe epub and pdf, and lrf). Other features that may or may not be found on specific e-readers include:

Content searchability

Content tagging (for creating a table of contents)

Organization of documents into folders, with related navigation

Search of folders and of the contents inside documents

Text-to-speech conversion for a few languages

Existence of third-party software development tools

An issue that is of concern to technical writers whose content will appear on e-readers involves reflowable versus nonreflowable content. In a word processor, text editor, or online publishing system, the content is said to be "reflowable" if it can automatically wrap words to the next line as the user changes the window size and thereby relocates the right margin of the page. A document that is not reflowable is fixed in its page layout and proportions. Changing those characteristics requires a redesign of the page at the point of creation, and can't be done "on the fly" by the e-reader device.

Reflowability is desirable for documents that are written for and will be read by e-readers. A reflowable document is available for download in a single file version, and reflowing is done locally and automatically by an e-reader device and is therefore efficient. In the nonreflowable approach, the document is available in as many files as the formats offered, depending on screen size and other factors, and the time needed to download multiple files can lead to slow performance.

Reflowable content requires fewer files to be stored for distribution (one file per document) and is an easier approach for the user (e.g., no misunderstanding as to which file to download). But with reflowable content there may be difficulties in resizing figures while maintaining readability across different screen resolutions and sizes. There is also a need to convert from "more popular formats," which may introduce layout and pagination errors.

9.7.2 Distribution Model

One last consideration for e-magazines is the distribution model. Electronic documents can be distributed with digital rights management (DRM), non-DRM, or as open access. DRM is intended to prevent unauthorized copy and distribution.

DRM ranges from "soft" to "hard" enforcement, where soft DRM is enforced with a simple watermarking mechanism and hard DRM limits certain operations (e.g., copying, printing, text extraction, reading aloud, annotating, and text-to-speech conversion). DRM schemes should, but do not always guarantee portability of content among devices belonging to the same user.

In non-DRM, the user has the obligation not to do unauthorized copy and distribution of the material.

In open access, the user is not only able to access free copy and distribution, but also to make modifications of original document, depending on the type of open access license used by the content's creators.

Current trends strongly depend on user history and experience using the various types of electronic products. For example, music started with hard DRM and later migrated to softer DRM (e.g., watermarking in iTunes). E-books generally use hard DRM, although the trend seems to be toward softer DRM.

9.8 Online Courses

Most universities, including mine, offer online courses and programs of study for technical professionals. Developing an online course is really a special kind of technical writing. While you may not be a university professor, many companies are putting product training and in-house training online, and you might be asked write one of these courses.

Here is some advice on writing online course from Fish and Wikersham [2009].

Consider the needs of the audience. Online students are typically adult learners. This means that it can be presumed that students are self-motivated, mature, and capable of independent learning. Still, keep the writing simple and understandable by applying all the best practices of technical writing that have been discussed in this book.

Provide a high level of organization and planning. It is common for online courses to simply be constructed by posting lecture notes and providing a few links to reading materials and Web sites. The best online courses have a more defined logical structure, and navigational guides providing links between related materials and links to external learning navigational guides.

Maintain a high level of student interaction and feedback. While this high level of feedback depends on the instructor, a well-designed online course creates opportunities frequent interaction. The best online courses have frequent interaction using threaded discussion forums, chat rooms, emails, and gradebook comments. Mid-course surveys to assess student satisfaction are also advisable.

Provide ongoing assessment. Provide some form of brief assessment such as a short quiz or writing assignment after each learning module. One or two true-or-false question "learning" checks are also useful in giving students a sense of confidence and creating an early warning system if something is wrong.

Incorporate more graphical elements and dynamic content. To keep the students' interest incorporate more pictures, graphics, videos, and simulations in the course. Online learning platforms provide support for these features and students expect to see these in courses. Recorded lectures have some value, but do not center the course around these.

I have been teaching about 70% of my courses online since 2010 and I have I've written several courses. My experiences are consistent with this advice.

9.8.1 Massively Open Online Course (MOOC)

A massively open online course (MOOC) is a free, open access online course that enrolls large numbers (sometimes) thousands of students in a single section. A MOOC has two basic models. The first involves Web-based and emailed course content, with assessment via automated exams. The second form, "connective" learning, has less structure and content. The learning presumably occurs via crowdsourced interactions through blogs, threaded discussion boards, and email. In either model, course assistants might moderate the interactions and answer questions, but instructor-initiated interaction are sporadic at best [Laplante 2013].

The financial model for MOOCs is still fuzzy. One model used by universities is to give the course away for free, but optionally charge for credits upon completing the course and passing all exams. Another model is for sponsors (for example, corporation) to pay for delivery of the course and then be allowed to recruit from the students who pass the course. Yet another method is to use the MOOC to sell related products (such as books, equipment, or services). Other financial models will eventually emerge, but at this writing, no entity has seemed to be able to profit from MOOC delivery.

9.8.2 Vignette: Experiences with MOOCs

To date, I've written three MOOCs. One is an introduction to cloud computing, which has had a least 100,000 students from more than 100 countries. The second course is an introduction to real-time operating systems, which has had less success with around 10,000 students. The third course was a specialized course about writing multiple choice exams for technical subjects. The audience for the last course was professors and teachers, and I don't know how many people have taken the course. Writing these courses was both challenging and fun. The real challenge involved distilling the course material into very short bites, recognizing that the attention span for MOOC students is relatively low (less than 5% who start a course actually finish it).

The first two MOOCs are easily found on the Web and I invite you to view them. The third course is only available through the IEEE.

9.9 Exercises

9.1 Write an e-mail to a colleague describing a technical problem that you recently experienced at either work or home.

9.2 Write an e-mail to a manufacturer of some technical product that you recently purchased describing any problems you had in setting up the product. If you had no such problems, describe the ease with which you set up the product.

9.3 For the following communities of interest, determine if they have a social media presence and by what mechanisms they communicate.

    a. Analog Computer Users

    b. Ugly Dog Fans

    c. Klingon Speakers

    d. Left-Handed Guitarists

    e. Beet Lovers

9.4 Write a simulated posting to a user group of your choice. Instead, you can write and post something to a real user group.

9.5 Write a simulated e-newsletter for a group of your choice.

9.6 Survey the Web for various e-reader devices. Construct a summary feature list for these devices. Represent your findings in the form of a table.

9.7 Do you belong to any professional or recreational communities online? If not, interview a friend who does belong to one. Then write a one page essay describing the characteristics of the community: who belongs, what are their interests, what happens in that community and so forth.

9.8 Have you ever taken an online course? If yes, what were your experiences? If no, find someone who has taken an online course and relate their experiences.

9.9 Find a MOOC on technical writing and list the topics. How are the topics in this MOOC different from those in this book?

9.10 Write a one page essay discussing the leading providers of MOOCs. Who are they? What business models do they use?

Endnotes

1. Tom Costello, CEO of Upstreme, www.upstreme.com.
2. We neither transferred copyright to the Web host nor to each other; therefore, Tom and I each own the copyright to our own postings.
3. Thanks to Professor Paolo Montuschi, University of Torino, for contributing to this discussion.
4. Thanks to Professor Paolo Montuschi, University of Torino, for contributing to this discussion.

References

Antón, C., Camarero, C., and Rodríguez, J., Pleasure in the use of new technologies: The case of e-book readers, *Online Information Review*, April 10, 41(2), 219–234, 2017.

Antón, C., Camarero, C. and Rodríguez, J., Usefulness, enjoyment, and self-image congruence: The adoption of e-Book readers. *Psychology &. Marketing*, 30, 372–384, 2013.

Derks, D., Bos, A.E.R., and Von Grumbkow, J., Emoticons in computer-mediated communication: Social motives and social context, *CyberPsychology and Behavior*, 11(1), 99–101, 2008.

Fish, W.W., and Wickersham, L.E., Best practices for online instructors: Reminders, *Quarterly Review of Distance Education*, 10(3), 279–284, 319–320, 2009.

Hemingway, M., Philly requiring bloggers to pay $300 for a business license, *Washington Examiner*, October 22, 2010, accessed November 2010, http://www.washingtonexaminer.com/philly-requiring-bloggers-to-pay-300-for-a-business-license/article/131121, accessed January 1, 2018.

Laplante, P.A., Courses for the masses?, *IT Professional*, March, 15(2), 57–59, 2013.

Laplante, P.A., IT predictions 2010, *IT Professional*, 12(1), 53–56, 2010.

Laplante, P. and Costello, T., Driving business value through IT during tough times, October 8, 2008, *IT World Open Exchange*, http://www.itworld.com/phil-laplante-and-tom-costello (This content is no longer available at the site.)

Lee, J., Is that an emoticon in 1862?, *New York Times*, January 19, 2009. http://cityroom.blogs.nytimes.com/2009/01/19/hfo-emoticon/, accessed January 1, 2018.

Milojicic, D., and Laplante, P., Special technical communities, *Computer*, June, 44(6): 84–88, 2011.

Wiesner, P., Laplante, P., Loeb, M., Gallus, T., Virtual communities and the technical professional, *Proceedings of the Frontiers in Education Conference*, 8(1), 2003, https://peer.asee.org/11505.pdf, accessed 1/3/2018.

Wikipedia, Emoticons, http://en.wikipedia.org/wiki/List_of_emoticons, accessed January 1, 2018.

10

Writing with Collaborators

10.1 Introduction

When writing substantial documents, and sometimes even brief ones, you will often have to work with others. In writing, as in any group endeavor, collaboration can be challenging. These challenges include noticeable and distracting variations in writing style, the mechanics of managing and tracking each other's changes, and occasionally negative interpersonal dynamics.

The severity of these problems increases as the team size grows because the number of interactions between people increases. In fact, the growth of the number of interactions is exponential. To see this, let n be the number of people in the group. The number of pairwise interactions is given by

$$\frac{n \cdot (n-1)}{2} = \frac{n^2 - n}{2} \tag{10.1}$$

Equation 10.1 can be visualized in Figure 10.1.

Encyclopedias, dictionaries, and contributed volumes may have hundreds of authors. But many technical books and papers may have two or more, sometimes dozens of authors. We would assume that these coauthors are colleagues and friends, but any one of these interactions can become toxic. When writing collaboratively, as with any group work, you will have to manage egos (including your own) and share credit, even when credit may not be due.

In a survey of coauthoring best practices, Noël and Robert found that the "most respondents used word processors and email for collaborative writing. Respondents noted the importance of change tracking, version control, and synchronization features. The study also found that there was no dominant form of group membership, project management, writing strategy, or task scheduling" [Noël and Robert 2004]. Wikis, a special kind of website with access control mechanisms and version control, may be employed to help manage the group writing process. But in the end, completing large collaborative writing projects is not about tools—it is about managing schedules, expectations, and egos, and that is the focus of Chapter 10.

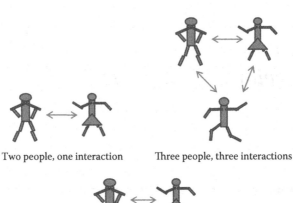

Two people, one interaction   Three people, three interactions

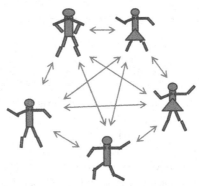

Five people, 10 interactions

FIGURE 10.1
The number of pairwise relationships in groups grows exponentially according to Equation 10.1.

10.2 Writing in Different Voices

One of the most common problems in collaborative writing involves a noticeable difference in writing style ("voice") from one section of the document to another. It is preferable that the writing appear to be by the same person throughout the document.[1]

Let's say that you and three coauthors are assigned to draft different parts of the document. Your plan is to assemble these pieces and make one or two final editing passes. But when you and your coauthors deliver your work and put it together, the result is terrible. Each section clearly looks like it was written by a different person. Important information is missing but some ideas appear duplicated in places. I call the problem of different voices the "writing homogenization problem." Let's see how you might solve this problem.

10.2.1 Using Metrics to Detect Nonhomogeneous Writing

In Chapter 3, I mentioned the Flesch–Kincaid metrics computed by Microsoft Word. These metrics can be computed for different sections of a draft document to see if a dramatic change in writing style occurs from one section to the next. If such a noticeable change is identified, then editing efforts can be directed toward homogenizing the writing.

Here is an example of what I mean. For a section of text I wrote for this book, Word computed the values for the Flesch–Kincaid statistics shown in Figure 10.2. The figure shows a Flesch Reading Ease of 40.3, and a Flesch–Kincaid Grade Level indicator of 12.1 (my writing tends to be in the Grade Level range of 10.5 to 12.5). Suppose another section of text had a Reading Ease of 25 and a Grade Level of 8. This suggests that the other section was not written by me and that this writing style difference would be noticeable to a reader. Therefore, editing of one or both sections is required. After editing, the reading statistics can be compared again. The cycle continues until the writing statistics are similar for all text sections, for example, within a 20% difference.

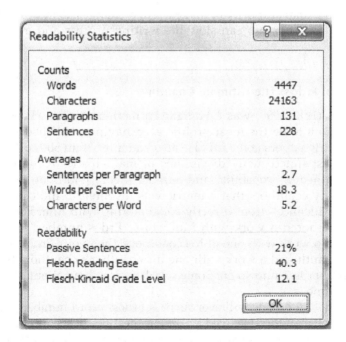

FIGURE 10.2
Readability statistics for a draft section of this book.

10.2.2 Dealing with Different Voices

Another possible attack on the "writing homogenization" problem is to have several rounds of circular revision. For example, suppose Fred, Sue, and Bob are collaborating, with each of them having lead writing responsibility for a different section of the document. After assembling the sections, Fred makes an editing pass through the entire document, then gives it to Sue for another round of revision and finally to Bob, who does likewise. Along the way, the reading scores are computed to see if the writing level is becoming uniform. If the writing is still nonuniform, then more editing passes through the entire document are made by each of the coauthors.

Another approach is for a person other than the collaborating authors to make one or more editing passes through the consolidated document. For example, in the case of Fred, Sue, and Bob, they might ask Emily to read and edit the draft document. This approach permits the authors to work on other projects while a fresh perspective is applied to the various disparate sections of the entire draft document.

I have found that you can achieve one voice in collaborative writing using any of these techniques, but the real key is for one of the coauthors to have final authority to make needed changes. A more democratic process of coauthorship can slow the writing process and even lead to stalemate.

10.2.3 Paul Erdős: The Ultimate Coauthor

Paul Erdős (1913–1996) was a Hungarian mathematician who is considered by many to be the most prolific ever, having published more than 1500 scholarly papers (a record). He also had more than 500 collaborators. His works spanned many disciplines of mathematics including graph theory, set theory, probability and number theory [Hoffman 1998]. Erdős had so many coauthors that a metric was invented to indicate the "collaboration distance" from directly collaborating with him. For example, if you coauthored a work with him, your "Erdős number" is 1. If you coauthored a work with one of his coauthors, then your Erdős number is 2. If you coauthored a work with one his coauthor's coauthors then your Erdős number is 3, and so on. Some scholars even brag about a low Erdős number (mine is 4).

It is likely that no one will ever surpass Erdős record number of publications and collaborators. He lived a peculiar lifestyle—it seems every hour of his waking life was devoted to mathematics. He essentially lived off of his friends and collaborators, had few positions and no home. He never married, and his only social activities, apparently, were centered on mathematical collaboration. He truly was a unique person.

10.3 Very Large Collaborative Writing Projects

Certain writing projects may involve dozens or even hundreds of writers, reviewers, and editors. My experience in this regard includes building dictionaries and encyclopedias. It is unlikely that you will be involved in these kinds of projects as a lead editor, but you may one day be invited to contribute to such a project. You might also be involved in some capacity in a large-scale project such as compiling procedure manuals or strategic planning documents. Therefore, I think it is worthwhile to describe my experiences with these kinds of projects and offer some advice.

10.3.1 Vignette: Building a Dictionary[2]

I have been the editor-in-chief for three dictionaries [Laplante 1998, 2001, 2005] and four encyclopedias [Laplante 2010, 2015, 2017, 2018]. These volumes are not so much writing or editing as they are management projects using the divide-and-conquer approach. But they provide a detailed case study of one way to manage large collaborative writing projects.

For example, consider how I built the *Dictionary of Computer Science, Engineering and Technology* [Laplante 2001]. The dictionary included material from more than a hundred contributors from seventeen countries. I knew this project would require a divide-and-conquer approach with fault tolerance. So, I partitioned the dictionary by defining areas that covered all aspects of computer science, computer engineering, and computer technology using the areas described in the *Denning Report on the Computer Science Curriculum* [Denning 1989]:

Algorithms, Data Structures, and Problems

Artificial Intelligence

Communications and Information Processing

Computer Engineering (Processors)

Computer Engineering (I/O and Storage)

Computer Graphics

Database Systems

Numerical Computing

Operating Systems

Programming Languages

Software Engineering

Robotics and Synthetic Environments

Computer Performance Analysis

I then recruited editors for each area from my network of friends and colleagues. Once the area editors were assigned, constructing the dictionary consisted of the following steps:

1. Creating a terms list for each area
2. Defining terms in each area
3. Cross-checking terms within areas
4. Cross-checking terms across areas
5. Compiling and proofing the terms and definitions
6. Reviewing the compiled dictionary
7. Final proofreading

The first task for the area editors was to develop a list of terms to be defined. A terms list is a list of terms (without definitions), proper names (such as important historical figures or companies), or acronyms relating to their areas. The content of each list was left to the discretion of the area editor based on the recommendations of the contributing authors. However, lists were to include all technical terms that related to the area (and sub-areas). Although the number of terms in each list varied somewhat, each area's terms list consisted of approximately 700 items.

Once the terms lists were created, they were merged and examined for any obvious omissions. These missing terms were then picked up from other sources with permission. The process of developing and collecting the terms took one and a half years. Once all the terms and their definitions were collected, the process of converting, merging, and editing of terms began. These steps took an additional six months.

Although authors were provided with a set of guidelines to define terms, they were free to exercise their own judgment and to use their own style. As a result, the entries varied widely in content from short, one-sentence definitions to rather long dissertations. While I tried to provide some homogeneity in the process of editing, I tried not to be heavy-handed in my editing and possibly corrupt the meaning of the definitions. Nor did I want to interfere with the individual styles of the authors. As a result, I think the dictionary contains a diverse and rich exposition that collectively provides good insights into the areas intended to be covered by the dictionary. Moreover, I was pleased to find the resultant collection much more lively, personal, and user friendly than typical dictionaries [Laplante 2001].

10.3.2 Vignette: Building an Encyclopedia[3]

Compiling an encyclopedia is another collaborative writing (or management) project type and is similar to assembling any large document from a variety of sources, such as product catalogues, group technical reports and major proposals, Consider the following experience when I was editor-in-chief for the *Encyclopedia of Software Engineering* [Laplante 2010].

The challenge of coordinating the activities of contributors and reviewers is similar for both encyclopedias and dictionaries; when I agreed to edit the *Encyclopedia of Software Engineering*, I thought I knew what I was doing.

The process started in January of 2007 with the formation of my editorial advisory board. Together we identified articles for each of the sub-areas corresponding to the knowledge areas of the IEEE/ACM's Software Engineering Body of Knowledge (SWEBOK) [IEEE 2004]. Then we identified and recruited expert authors to write these articles. This step was not easy. Teasing small entries for a dictionary is much easier than extracting substantial articles for an encyclopedia, and experts are always busy. There were many false starts and stops, searches for new authors when necessary, and the constant need to offer encouragement to the volunteer contributors. In the end, more than 200 expert contributors and reviewers from the industry and academia from twenty-one countries were involved.

As soon as contributors sent in first drafts, I had to send them out for peer review. Finding expert peer reviewers, who are also busy, is not an easy task.

When peer reviews were complete, the articles and review reports were returned to the authors for revision, and in many cases, another round of reviews. The effort was similar to editing a special issue of a scholarly journal, only magnified by a factor of twenty.

The final articles then needed to be edited by expert copyeditors, then returned to the authors for another check. I conducted one final check. It should come as no surprise, then, that the process of creating an encyclopedia from start to finish took nearly four years.

Completing the encyclopedia made me appreciate the effort in building the first *Oxford English Dictionary (OED)*, as told in the *The Professor and the Madman* [Winchester 2005]. The OED took forty years to complete without relying on enabling technologies such as the Internet. As it turned out, the most prolific contributor to that volume was actually an inmate in an asylum for the criminally insane. Many times while editing the encyclopedia, I could not tell if I was the professor or the madman. Since completing my first encyclopedia I have assembled several others [Laplante 2015, 2017, 2018] and each successive one was harder to complete. It seems that finding good writers who are willing to contribute articles is increasingly difficult.

10.4 Behavior of Groups[4]

When dealing with teams of writers, editors, or technology professionals, it is good to understand that newly formed groups do not always function effectively at the outset. Setting reasonable expectations of how much can be accomplished quickly by newly formed teams helps in making a realistic writing schedule and also in helping to steer the team's development in a positive manner through positive and reinforcing behaviors.

10.4.1 Tuckman's Model

Bruce Tuckman's classic work provides a basic understanding of team formation [Tuckman 1965]. This highly cited and still relevant work was the first to describe the life cycle of groups from potentially chaotic mobs to well-functioning teams. Tuckman's model will be described in the context of a large writing project consisting of writers, editors, managers, and other persons who have some role. Typical large writing projects include requirements documents, design documents, strategic plans, and user manuals.

Tuckman contends that teams can dramatically change from one form to another over time. This evolution may be gradual and therefore somewhat imperceptible to the members. Tuckman described the signs of development as evolving through a development life cycle, which is often characterized as "forming, norming, storming, performing, and mourning" (Figure 10.3).

Tuckman's "model" is just that—a model. It is simply a way to try to characterize a group's transformation, and the timeline for each team will be different. Moreover, the transition from one phase to another won't necessarily be discrete. Let's look at each of these phases in more detail.

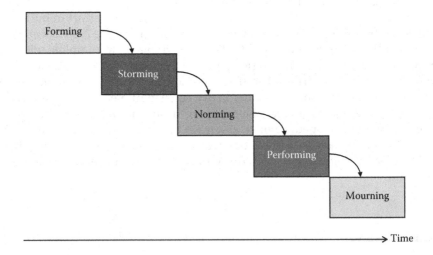

FIGURE 10.3
Tuckman's phases of team development. (From Tuckman, B., *Psychological Bulletin*, 63, 384–399, 1965.)

10.4.2 Forming

Formation occurs when the writing team is initially composed. During this period, members of the team do the usual sizing up of one other and the testing of personal boundaries. In the forming stage, there is usually confusion about direction, and little work gets accomplished. While this situation is considered "normal," team members who are eager to move forward may feel temporarily frustrated.

10.4.3 Storming

In the storming phase, group members recognize that the group needs to begin to evolve into "something," although that "something"—that ideal vision of the group—may be different for each member. These differing visions of what the team should be and do, even in the presence of a stated or assigned vision, is often the cause of great tension, hidden agendas, and anger. The storming phase is perhaps the most difficult stage for the team because team members begin to realize that the road ahead is different and more difficult than they had expected. Some team members become impatient with the slow progress and there is disagreement about next steps. Team members are not yet comfortable working in the group, so they continue to rely on their own personal experience and skills. At this stage, they resist collaborating with their fellow members.

In the environment that creates these feelings, the pressures are such that little time is actually spent on writing and problem solving. With the aid of a good leader, however, a team can progress rapidly through this phase. Without a good leader, the team can languish.

10.4.4 Norming

At some point in the evolution, a majority of the team recognizes that it must escape the storming, although there may not necessarily be an explicit recognition that "we are in the storming phase." Characteristically, the team begins to converge on a shared vision, or at least a shared set of tasks. Ideally, the team may even begin to set its sights beyond the original scope of the task, choosing goals that are more ambitious. Enthusiasm is high and, more importantly, most members begin to set aside their petty differences and agendas, or at least subordinate them. The majority of the team begins to accept the rules of the group and the roles that have been assigned to them. A noticeable reduction in tension occurs as trust gradually wins out over ego, although there still may be some flare-ups. Nonetheless, healthy disagreement dominates over nonproductive argument.

As a result of the normative behavior, the team begins working on the tasks at hand. It is sometime during this phase that noticeable progress toward completion of the writing goals will be seen.

10.4.5 Performing

At the performing stage, the team members have finally settled their relationships and expectations. Team members have discovered and accepted each other's strengths and weaknesses, and have identified their roles. Their efforts are now directed toward diagnosing and solving problems, and choosing and implementing real changes.

The team is now an effective, cohesive unit. The sign of a performing team is that real writing is getting done regularly. At this point, managers should use most of their energies to maintain this state of team function, and to see that it is not upset.

10.4.6 Mourning

Finally, during the mourning (or "adjournment") phase, the team debriefs and shares the improved process. When the team finally completes that last briefing, there is always a bittersweet sense of accomplishment coupled with the reluctance to say good-bye. Many relationships formed within these teams continue long after the team disbands.

Tuckman's model is mostly useful to help us understand that team formation is not easy, and to recognize the signs when the team seems stuck in one of the phases prior to "performing." While the Tuckman model suggests that teams will eventually reach the performing stage, it is dangerous to assume that this will happen naturally and predictably. Without effective leadership, the team can get stuck in a suboptimal phase, or at least be delayed in its progression. The progression may happen of its own accord or, more deliberately, may be guided and accelerated by proactive team members and managers.

10.4.7 Vignette: Determining Author Order

Whether your name appears as the first author on a coauthored publication can matter. In most academic disciplines and in industrial research, there is an implication that the first author listed contributed the most and that the further in the order that an author is listed, the less they contributed to the work. It may seem petty, but in academia, particularly in certain disciplines such as mathematics and engineering, author order is quite important. For example, a promotion or tenure can be denied if the candidate has rarely been first author on a paper, has had too many papers where she is "lost" amongst numerous coauthors or has never had a solo authored publication.

The situation gets murkier in those scientific disciplines where it is customary to list everyone working in a laboratory as a coauthor of any published work, regardless of contribution, Hence, for certain papers, say in chemistry or physics, there may be ten or more coauthors. Sometimes feuds occur over "first author" status.

Here is an apocryphal situation that illustrates the importance of fairness when assigning authorship. A productive semiconductor research laboratory, with more than 20 members, was publishing dozens of high quality papers per year. But the question of author order, particularly who would be first, was always contentious.

Trying to resolve the problem, the lab members held a summit meeting. One member suggested a random assignment of authors. Another advocated that each member should take turns as first author and the remaining author order determined randomly. A third member proposed using a voting system to determine author status. The head of laboratory, we'll call him Dr. Arthur Aardvark, had a different idea. He was responsible for administrative duties, providing financial support, hiring staff, and external relationship building. He contributed very little to the actual research activities. So, when he proposed listing the authors alphabetically, which would have made him the first author of all research papers, the other lab members objected. But as he was the head of the laboratory, his suggestion prevailed. Within two years, the lab disbanded because of low morale and high employee turnover.

10.5 Other Paradigms for Team Building[5]

10.5.1 Group Writing and Improvisational Comedy

Improvisational comedy can provide some techniques for collaboration and for dealing with adversity in group writing projects. Anyone who has ever enjoyed improvisational comedy (e.g., the British and American television show *Whose Line Is It Anyway?*) has heard the exquisite verbal interplay of people with very different points of view and the resolution of those differences.

I have been a fan of improvisational comedy for quite some time (it is a useful skill for a college professor), and I think there are several lessons that can be taken away from that art:

Listening skills are really important, both to hear what others are saying and to play off your partner(s) in the collaborative writing effort.

When there is disagreement or partial agreement, the best response is "Yes, and…" rather than "No," or "Yes, but…" That is, build on, rather than tear down ideas.

Things will go wrong—both improvisational comedy and collaboration are about compromise.

You should have fun in the face of adversity.

Finally, you should learn to react by only controlling that which is controllable. Usually, you can only control your own reaction to certain events, not the events themselves—and certainly not your collaborators' reactions.

You can practice some techniques from improvisational comedy to help you develop your listening skills, emotional intelligence, and ability to think on your feet, which in turn will improve your ability to collaborate in writing and other projects.

For example, consider one improvisational skill-building game called "Zip, Zap, Zup (or Zot)."[6] Here is how it works. Organize four or more people (the more, the better) to stand in a circle facing inside. One person starts off by saying zip, zap, or zup. If that person looks at you, you look at someone else in the circle and reply in turn with one word—zip, zap, or zup. Participants are allowed no other verbal communication than saying one of the three words. The game continues until all participants begin to anticipate which of the three responses is going to be given. The game is a lot harder to play than it seems, and the ability to anticipate responses can take several minutes (if it is attained at all). The point of this game is that it forces participants to "hear" emotions and pick up on other nonverbal cues.

In another game, called "Dr. Know-It-All," three or more participants answer questions together, with each participant providing just one word of the answer at a time. So, in a technical writing exercise, we would gather a collection of writers and have them write out loud, meaning that each person in the group provides one word, followed by the next person, in round-robin fashion. This is a very difficult exercise, and it is not intended as a writing technique. It is a thought-building and team-building exercise, and it can help to inform the participating writing collaborators about the challenges that lay ahead.

One final exercise involves answering questions from two other persons at the same time. This experience helps participants think on their feet. It also simulates what they will often experience when interacting with customers or collaborators where they may need to respond simultaneously to questions from two different people.

It seems clear that our brains tend to suppress our best ideas, particularly under stress. Improvisation helps us think spontaneously and creatively. So, try these exercises for fun while developing these important skills.

10.5.2 Team Technical Writing as Scriptwriting

The writing of screenplays for movies has quite a few similarities to group writing endeavors. Norden describes how requirements engineers can learn how to resolve different viewpoints by observing the screenwriting process, and these observations can be adapted to group writing of technical materials [Norden 2007].

There are many similarities between the making of movies and collaborative technical writing. Of course, both are efforts involving more than one person, but there are more profound similarities. For example, movies are announced well in advance in order to build excitement, but these expectations may not be met when the movie is delivered (e.g., changes in actors, screenwriters, directors, plots, and release date). Technical writing projects are often announced in advance of the actual publication date, which may not be met, and may promise features that do not materialize. Egos (including your own) are often a significant factor in both movies and technical writing.

Movies are shot out of sequence and then assembled to make sense later. Technical writing is often constructed "out of order," too; that is, the parts are written nonsequentially. A great deal of work ends up getting thrown away—in movies it ends up as scenes that are cut[7] and discarded during editing; in technical writing it is text that gets edited out of the final version.

What can technical writers learn from big-picture production? In a short vignette within Norden's article, Sara Jones provides the following tips for requirements engineers, learned from screenwriting, with the idea that they can be adapted for technical writing.

> Preparation is everything. Don't leap straight into writing without doing your research.
>
> Expect to work through many drafts.
>
> Think in detail about the readers of the document being prepared.
>
> Hold your audience—remember that someone must read whatever it is that you are writing. Try to make your writing compelling [Norden 2007].

I have modified this list slightly, to reflect my experience in collaborative writing projects.

10.6 Antipatterns in Organizations[8]

When collaborating on a writing team, many challenges can arise from organizational dysfunction. These challenges include difficulties in communication, resource acquisition and sharing, jealousy, and lack of recognition for successes. But it is not always easy to identify and then "treat" organizational dysfunction. When problems are correctly identified, they can almost always be dealt with appropriately. But organizational inertia frequently clouds the situation or makes it easier to do the wrong thing rather than the right thing. So how can you know what the right thing is if you have not accurately identified the dysfunction?

Organizational dysfunctions often appear as "antipatterns." An antipattern is a recurrent problem form, that is, it can appear in any organization and the symptoms (and potential cures) are nearly identical every time. Antipatterns can bubble up from the individual employee or manager through organizational dysfunction and can manifest as badly stated, incomplete, or incorrect rules and procedures. Even worse, they can evolve into intentionally disruptive environments. Using antipatterns to model and solve organizational problems assists in the rapid and correct identification of problem situations, provides a playbook for addressing the problems, and provides some relief to the beleaguered employees in these situations in that they can take consolation in the fact that they are not the first people to face the same issues. It is beyond our scope to discuss why antipatterns arise. But some antipatterns are the result of misguided corporate strategy or uncontrolled social or political forces.

In dealing with the challenge of writing in large collaborative projects, I will discuss two likely environmental antipatterns: "divergent goals" and "process clash." These antipatterns can also manifest as differences in styles between two persons (see Vignette 10.6.3 Experiences Co-Writing Books and Papers). For a thorough discussion of other antipatterns found in organizations, particularly those that emerge in software development, see Laplante and Neill [2006].

10.6.1 Divergent Goals

Everyone works toward the same objectives. There is no room for individual or hidden agendas that don't align with those of the business. The divergent goals antipattern exists when people work toward opposing objectives.

There are several direct and indirect problems with divergent goals:

Hidden and personal agendas divergent to the mission of an organization starve resources from strategically important tasks.

Organizations become fractured as cliques form to promote their own self-interests.

Decisions are second-guessed and subject to "review by the replay official" as staff try to decipher genuine motives for edicts and changes.

Strategic goals are hard enough to attain when everyone is working toward them. Without complete support, they become impossible to achieve.

There is a strong correspondence between stakeholder dissonance and divergent goals, so be very aware of the existence of both.

Because divergent goals can arise both accidentally and intentionally, there are two sets of solutions or refactorings.

Dealing first with the problem of *comprehension and communication* involves explaining the impact of day-to-day decisions on the organization's larger objectives. Misunderstanding may be due to the fact that individuals don't understand that their decisions have a broader impact on the organizational mission and goals than they realize. Certain team members may have a very narrow perspective on their role or area within the organization, and that must be broadened. In this case, education or training coming from those with a wider (or longer-term) organizational perspective can help folks better understand how their seemingly localized decision making adversely affects the larger mission.

The second problem of *intentionally charting an opposing course* is far more insidious, however. Its remediation requires considerable intervention and oversight. The starting point is to recognize the disconnect between an individual's personal goals and those of the organization. Why do some people feel that the organizational goals are incorrect? If the motives really are personal—if the malcontent feels his personal success cannot coincide with the success of the organization—then radical changes are needed. Otherwise, the best recourse is to get the unhappy person to buy into the organizational goals. This objective is most easily achieved if every stakeholder is represented in the definition and dissemination of the core mission and goals, and subsequently kept informed, updated, and represented.

10.6.2 Process Clash

A *"process clash"* is the friction that can arise when advocates of different processes must work together without a proven hybrid process being defined. The dysfunction appears when organizations have two or more, well-intended but noncomplementary processes. In this situation, a great deal of discomfort can arise for those involved. Symptoms of this antipattern include poor communications, even hostility, high turnover, and low productivity.

One solution to a process clash is to train and cross-train the individuals with differing process sets. Another solution is to select an altogether different process that resolves the conflict.

10.6.3 Vignette: Experiences Co-Writing Books and Papers

I have wide ranging experiences with my numerous co-authors of books and papers. Not counting the hundreds of contributors to my dictionaries and encyclopedias, I have had more than 125 different co-authors—certainly no threat to Erdős' record but even I was surprised at how many when I counted. The majority of my collaborators have been fun and cooperative (characteristics I select for). But I have had some bad experiences too.

Failure to deliver promised work on time or at all is the most frustrating problem I have encountered. Sometimes a coauthor and I will disagree on some fundamental aspect of the research (and subsequent writing), in other words or goals diverge. Another common problem is discovering my coauthor and I work in way that is incompatible (a kind of process clash).

I have learned that the challenges of large writing projects are manageable if you approach them with a positive attitude and a willingness to compromise. In handling these negative situations with coauthors, I use some of the refactoring techniques discussed in Section 10.6. But I generally will not work again with someone who disappoints me.

10.7 Exercises

10.1 Write a two-page essay on the life of Paul Erdős.

10.2 If you were working in a situation (e.g., work or a university) where a group of collaborators produced a technical work for publication, how would you suggest determining author order?

10.3 If you can, describe a large collaborative writing project in which you have been involved. What challenges did you face? Did you overcome these challenges? If so, how? If not why not?

10.4 For a sample of your writing, use Microsoft Word or another program to compute the Flesch–Kincaid metrics. How do your metrics compare with those of a colleague?

10.5 Play the "Zip, Zap, Zup" game with a group of friends, colleagues, or classmates (no drinking allowed).

10.6 Use the Dr. Know-It-All game with a group of friends, colleagues, or classmates to write a short story (100 words or less) of your choosing.

10.7 Organize and play the "the two question game" (see Section 10.5.1).

10.8 Take the antipatterns test at https://phil.laplante.io/antipatterns .php What did you find?

10.9 How would you deal with a coauthor (of some technical writing) who is underperforming in some way? Discuss a real or hypothetical situation and how you dealt (or would deal) with it?

10.10 There are many examples of famous collaborators (for example, composers, actors, researchers) whose collaborations ended in spectacular failure and misunderstanding. Find an example of one of these and discuss the situation in one page essay. Conclude with suggesting how the collaborators could have resolved their differences.

Endnotes

1. An exception is a contributed book with each chapter written by a different expert.
2. Some of this section is excerpted and adapted from Laplante [2005].
3. This discussion is excerpted and adapted from Laplante [2010] with permission.
4. This discussion is excerpted and adapted from Laplante and Neill [2006] with permission.
5. This discussion is excerpted and adapted from Laplante [2009] with permission.
6. I know that this is also a traditional "drinking" game.
7. The film stock isn't physically "cut" anymore; all the editing is done to the digitized film source.
8. This discussion is excerpted and adapted from Laplante [2006] with permission.

References

Denning, P. (Ed.), Computing as a discipline, *IEEE Computer Journal*, 22, 63–70, 1989.

Hoffman, P., *The Man Who Loved Only Numbers*, New York: Hyperioncop, 1998.

IEEE (Institute of Electrical and Electronics Engineers, Association for Computing Machinery), Software Engineering Body of Knowledge, 2004, available at http://www.swebok.org/, accessed January 7, 2018.

Laplante, P. A. (Ed.), *Comprehensive Dictionary of Electrical Engineering*, CRC Press, Boca Raton, FL, 1998.

Laplante, P. A. (Ed.), *Comprehensive Dictionary of Computer Science, Engineering and Technology*, CRC Press, Boca Raton, FL, 2001.

Laplante, P. A. (Ed.), *Comprehensive Dictionary of Electrical Engineering, Second Edition*, CRC Press, Boca Raton, FL, 2005.

Laplante, P. A., *Requirements Engineering for Software and Systems*, CRC/Taylor & Francis, Boca Raton, FL, 2009.

Laplante, P. A. (Ed.), *Encyclopedia of Software Engineering*, Taylor & Francis, Boca Raton, FL, 2010.

Laplante, P. A., (Ed.), *Encyclopedia of Information Systems and Technology*, CRC Press, Boca Raton, FL, 2015.

Laplante, P. A., (Ed.), *Encyclopedia of Computer Science and Technology*, CRC Press, Boca Raton, FL, 2017.

Laplante, P. A., (Ed.), *Encyclopedia of Image Processing*, CRC Press, Boca Raton, FL, 2018.

Laplante, P. A. and Neill, C. J., *Antipatterns: Identification, Refactoring, and Management*, Auerbach Press, New York, 2006.

Noël, S., Robert, J. M. Empirical study on collaborative writing: What do co-authors do, use, and like?, *Proceedings of Computer Supported Cooperative Work Conference*, March 27, 13(1), 63–89, 2004.

Norden, B., Screenwriting for requirements engineers, *Software*, 24(4), 26–27, 2007.

Tuckman, B., Developmental sequence in small groups, *Psychological Bulletin*, 63, 384–399, 1965.

Winchester, S., *The Professor and the Madman: A Tale of Murder, Insanity, and the Making of the Oxford English Dictionary*, Harper Perennial, New York, 2005.

Appendix A

Writing Checklist

- Collect writing materials
- Brainstorm
- Write first draft
- Revise and rewrite several times
    - Combine two or more words to one powerful word
    - Check precision
    - Remove emotional indicators
    - Remove clichés
    - Fix malapropisms and erroneous homonyms
- 5 Cs check
- Spell check
- Copyright check
- Reference formatting
- Plagiarism check
- Final review and revision

Permissions Request

Dear Copyrights and Permissions Representative:

I am preparing a [book | paper | article] entitled, "<title>" to be published by <publisher> in <date>. I would like to use the following: <list the figure, table and/or text to be used>. I would like to use the material(s) in the following way <describe how the figure, table and/or text are to be used>.

Please advise if any additional information is required to fulfill this request.

Thank you very much,

<Your name>

Appendix B: Templates

Technical Article or Book Review
Review of: <Title of Manuscript> by <Authors> by <Your name—will be removed if blind review>

Confidential Comments to the Editor

<Provide any information that could help the editor make a decision about the paper but would not be appropriate for the authors to see. For example, if you think parts of the paper may have been plagiarized.>

Confidential Comments to the Authors

<Describe the contributions of the paper and its shortcomings. Suggest ways that the paper could be improved. Make recommendations about the references, for example, if important ones are missing or if there are unnecessary ones.>

Final Recommendation

<Accept paper as is | Accept paper with minor modifications | Require major revisions and re-review | Reject paper | Paper is out of scope>

Résumé

Name
Contact information

Summary

<50 word synopsis of your resume>

Statement of Objective

<Describe career objective>

Experience

<List your previous positions from most recent to oldest. Include name of company, your title, dates of employment. Describe duties, responsibilities and major achievements.>

Education/Training

<List education from highest level to high school. List any other training.>

Other

<Include other section headings as appropriate. See Chapter 5, Section 5.2>

References

<List name, position, relationship (e.g., supervisor, peer, subordinate) for 3–5 people who can discuss your work in detail.>

Letter of Reference

<Recipient's address>

<Your address>

<Date>

Dear <Recipient>

I am writing this letter of support for <candidate>. I have known <candidate> for <number> years in the following capacity <describe how you know the candidate>.

<Discuss your knowledge of candidate's specific accomplishments. Discuss positive aspects of candidate's demeanor and personality. Provide any anecdotes that are indicative of the candidate's strengths. Describe any areas where you think the candidate could grow; position this as an opportunity not a criticism.>

In summary <highlight key positive aspects of candidate>. I therefore give <him | her> my strongest recommendation without any reservation. Please do not hesitate to contact me if you have any questions.

Sincerely,

<Your name>
<Your title, if applicable>

<Phone number and email>

Agenda

AGENDA
<Name of Organization.
<Date and Time>

Welcome and introductions	<Person>	(Start time–end time)
<Topic 1>	<Person>	(Start time–end time)
<Topic 2>	<Person>	(Start time–end time)
<Topic 3>	<Person>	(Start time–end time)
<Topic 4>	<Person>	(Start time–end time)
...		
<Topic n>	<Person>	(Start time–end time)
Other business	<All>	(Start time–end-time)
Adjournment		(Time)

Next meeting: <Date and time>

Meeting Minutes

Attendees
<List all meeting attendees>

Topic 1
<Summarize discussions and key points pertaining to Topic 1.>

Topic 2
<Summarize discussions and key points pertaining to Topic 2.>

Topic 3
<Summarize discussions and key points pertaining to Topic 3.>

Topic 4
<Summarize discussions and key points pertaining to Topic 4.>

...

Topic n
<Summarize discussions and key points pertaining to Topic n.>

Other Business
<Summarize all other items raised and discussed.>

Action items
<List all action items created and responsible persons.>

Adjournment
<List the time the meeting adjourned.>

RECIPE

Name of Recipe
Serves <Number>

Ingredients

<Ingredient 1>
<Ingredient 2>
<Ingredient 3>
...
<Ingredient n>

Preparation

<Step 1>
<Step 2>
<Step 3>
...
<Step m>

Footnotes
<Any incidental, historical, or personal information>

Research Grant Proposal

Title: <Descriptive Title of Proposed Research>

Practices Problem Statement:
<Describe the problem that the research is supposed to solve or address.>

Hypotheses/Objective:
<Describe how your research solves this problem, makes progress towards its solution, or furthers the understanding of the problem.>

Scope and Limitations:
<Describe what is to be studied and any exclusions. List all limits to size or source of sample populations or specimens to be studied.>

Definition of Terms:
<Provide a glossary of domain-specific terms that are used in the grant proposal.>

Background:
<Provide an historical review of the problem and prior attempts to understand and solve the problem. Provide very accurate references to all works related to problem and your research>

Uniqueness of the Research:
<Describe how your approach is different or better or enhances any other prior works. Provide very accurate references to all works. This is the most important section of the proposal>

Potential Contribution:
<Describe why this research is important and how the results will impact society.>

Direct Application:
<Describe how the result of the research could be implemented immediately or in the future. Omit this section if this is purely theoretical (basic) research.>

Potential for Technology Transfer:
<Can the research be productized? If this is military research, are there civilian applications? Omit this section if this is purely theoretical (basic) research.>

Methods and Procedures:
<Describe how the research will be conducted in a stepwise manner. Describe how the results will be disseminated.>

Success Criteria:
<Describe how it will be determined if the research is a success.>

Budget:
<Provide a detailed budget. The budget should include funding for personnel, equipment, software, travel, overhead and any other related expenses.>

References:
<List all references to the discussions above. Use the reference style required or choose one if allowed. Do not list extraneous references. Always be consistent in referencing style.>

Consulting Services Proposal[1]

PROPOSAL
<Descriptive Title>
<Date>

<Your name> (Consultant) proposes to offer consulting services (Engagement) to <client name> (Client) at <location where consulting is to be offered>.

1. **Preamble.** The responsibilities of both the Consultant and Client are outlined in this Agreement. Each has obligations to one another, which when fulfilled in an atmosphere of mutual respect and cooperation, will yield benefits to all concerned.
2. **Work to be Performed.** <Summarize the work to be performed including all deliverables and dates of delivery.>
3. **Client Obligations.** <List all items to be furnished by client, for example, access to facilities, documentation, products, and personnel. List any insurance or licenses that need to be obtained by Client.>
4. **Consultant Obligations.** <List all materials to be delivered or returned upon completion. List any insurance or licenses that need to be obtained by Consultant.>
5. **Payment Terms.** <Describe how much is to be paid and when. List any advances to be rendered. Describe in what form the payments should be made (e.g., check, wire transfer, or other means). Discuss is money is due before, during or after the work or any partial payments. Note if there is a reward for fast payment or penalty for delayed payment.>
6. **Changes or Cancellation.** <Describe what happens if either the Client or Consultant cancels the contract. Discuss mechanisms for rescheduling the engagement if appropriate. Note if there is any penalty or partial payment due.>
7. **Interpretation** The services described this Agreement constitute the entire agreement between the parties hereto, and supersede all prior verbal or written discussions and agreements. This Agreement shall be construed in accordance with the laws of the Commonwealth of Pennsylvania and shall be deemed to have been accepted in said state. It may not be changed orally. Any controversy arising out of or relating to this Agreement or the breach thereof shall be settled by arbitration in <State/Province/Territory/Jurisdiction> in accordance with the rules of <any recognized arbitration authority such as the American Arbitration Association>, and the award rendered by the arbitrator may be entered in any court having jurisdiction thereof.[2]

Strategic Plan

Executive Summary

Mission Statement

SWOT Analysis

Competitive Market Analysis

Goal 1.
Objective 1.1
Objective 1.2
…
Goal 2.
Objective 2.1
Objective 2.2
…
Goal 3.
Objective 3.1
Objective 3.2
…
…
Goal n.
Objective n.1
Objective n.2
…
Appendix A. Strategic performance indicators

Appendix B. Budget

Other Appendices as needed

User Manual

Cover page

Quick start guide
  <If applicable>
Table of Contents

List of Figures

Installation/Assembly
  <If applicable>

Feature 1
  <Describe feature 1 in detail>

Feature 2
  <Describe feature 2 in detail>
...

Feature n
  <Describe feature n in detail>

Maintenance
        <Describe product maintenance>

Frequently Asked Questions
        <List FAQs and answers>

Troubleshooting
  <Provide step by step guidance to identify and fix common problems>

Questions or Problems?
  <Provide contact information>

Glossary

Index

E-Newsletter

<TITLE>

<Date><Issue Number>

WELCOME NEW MEMBERS
<List names of all new members>

ANNOUNCEMENTS
        ...
CONTRIBUTED ARTICLE 1
<Article 1>
        ...

CONTRIBUTED ARTICLE n
<Article n>
        ...

EVENTS
<List upcoming events>

    ...

MISCELLANEOUS
<Personal announcements, e.g., retirements, birthdays, marriages, deaths>
<Public service announcements>
< Advertisements>

...

Endnotes

1. For discussion purposes only. Always consult with an attorney before submitting any proposal for services.
2. I prefer arbitration instead of mediation or adjudication in a court of law in the case of a dispute.

Glossary

Abstract: (1) A prospectus of what a presenter intends to present at a conference. (2) A summary of a technical paper.

Administrative rejection: When a submitted article is rejected without formal review in a journal or magazine.

Allegory: A story that uses metaphors for real characters and events.

Anthropomorphic writing: Projecting human feelings, behaviors, or characteristics upon animals, inanimate objects, or systems.

Antipattern: A recurrent problem in organizations due to mismanagement or negative environment.

Authority: Refers to the reliability of the scientific content or to the qualifications of the writer.

Bethesda Statement: A statement of the principles of open access publication written at a conference in Bethesda, Maryland (USA).

Biosketch: A short biography used for various business purposes.

Blog: A website that features short, timely, and informal information snippets. "Blog" is a mash-up of the words "Web" and "log."

Bootleg book: A book that has been illegally reproduced and sold without the permission of the author and/or publisher.

Brainstorming: The process of recording your ideas on paper. Sometimes called "pre-writing."

Business communications: Any correspondence that must be written in the course of business activities.

Camera-ready: Material provided to a publisher that is ready to be printed without further editing or formatting.

Changeable: The structure of the document that will readily yield to modification.

Cite: To make reference to another work.

Clarity: When each sentence, related groups of sentences, or related sections of the written document can have only one interpretation.

Community of interest: A group with a shared focus, whether technical professional, political, recreational, religious, or other.

Completeness: No missing relevant or important information.

Concept map: A hierarchical organization of ideas from a central concept to various subconcepts and sub-subconcepts. Also called a mind map.

Conference: A meeting where researchers present scientific findings, often in preliminary form.

Consistency: In writing, if one part of the document does not contradict another part. (*See also* internal consistency and external consistency.)

Copyleft: Open source software or documentation the licenses that permit free use and redistribution of intellectual property provided certain rules are followed.

Copyright: The ownership rights for intellectual property.

Correctness: When the information is grammatically and technically correct.

Cover letter: (*See* transmittal letter.)

Curriculum vitae: A résumé for a professor or academic administrator. Commonly abbreviated as CV.

CV: (*See* curriculum vitae.)

Digital archive: A collection of published works that are placed in an electronically accessible, Web-based library.

Digital book: (*See* electronic book.)

Digital magazine: (*See* electronic magazine.)

Digital rights management: The type of distribution rights allowed for digital content. Also called DRM. (*See also* soft DRM and hard DRM.)

DRM: (*See* digital rights management.)

E-book: (*See* electronic book.)

E-newsletter: A simple periodical that is distributed exclusively via an e-mail server or downloaded from the Web.

E-reader: (*See* Electronic reader.)

E-zine: (*See* Electronic magazine.)

Electronic book: An electronic file containing a book and intended to be read on an electronic reader. Also called an e-book or digital book.

Electronic magazine: The equivalent of a printed, glossy technical trade magazine or refereed magazine, only in electronic format. Also called an e-zine.

Electronic reader: Electronic device designed for reading digital books and periodicals available in electronic form. Also called an e-reader.

Emoticons: Text-based graphics that depict facial expressions in order to represent an underlying emotion.

Existential: Use of the words "exists" or "none" and their equivalents. (*See also* universal quantification.)

External balance: When the relative number of major and minor sections and subsections is relatively uniform. (*See also* internal balance and hierarchical writing.)

External consistency: When a document is in agreement with all other applicable documents and standards. (*See also* internal consistency.)

Flesch–Kincaid metrics: Reading ease and grade-level indicators that are computed for writing by various word processors.

First-person point of view: Writing that is from the point of view of the author. (*See also* second-person writing and third-person writing.)

Formal method: A system of rigorous semantics used for the representation of documentation, such as requirements specifications. Formal languages look like a combination of a programming language and mathematics.

Freelance writer: A professional writer who works as an independent consultant serving many clients.

Glossary: A list of terms and their definitions, proper names (such as important agencies, organizations, or companies), and acronyms relating to the subject at hand.

Gold open access: When authors pay a fee to make their work publicly accessible through an open access publisher.

Green open access: When authors freely make their work publicly accessible through an open access publisher.

Hard DRM: A type of digital rights management for electronic content in which certain operations are restricted, such as copying, printing, and text extraction. (*See also* soft DRM.)

Heterographs: Two words that sound the same but are spelled differently and have different meaning, for example, "their" and "there."

Hierarchical writing: Writing that is arranged as a cascade of sections or chapters at a high level of abstraction, followed by sections and subsections of increasing level of detail. (*See also* external balance and internal balance.)

Intellectual property: Pertaining to a broad class of content produced by an author or artist. Any correspondence, article, book, figure, drawing, photograph, song, poem, etc. is intellectual property. Commonly abbreviated as IP.

Internal balance: When the relative lengths of the sections and subsections are uniform. (*See also* external balance and hierarchical writing.)

Internal consistency: When one part of the document does not contradict another part. (*See also* External consistency.)

IP: (*See* Intellectual property.)

Libel: In writing, a defamatory or deliberately false statement intended to injure the reputation of a person.

Malapropism: A word that sounds similar to an intended word but is logically wrong.

Market analysis: A report that examines the market demographics, customer characteristics, economic trends, and competition for some organization.

Mind map: (*See* Concept map.)

Minutes: The written record of a meeting.

Monograph: A book usually sole authored on a highly esoteric and advanced technical subject.

MOOC: Massively open online course, a free online course that can have thousands of students enrolled.

News release: (*See* Press release.)

Newsletter: An informal publication produced by some community of interest.

Nonreflowable: In a word processor, text editor, or online publishing system, when the document is available in as many files as the formats offered, depending on screen size and other factors. (*Contrast with* reflowable.)

Open access publication: An intellectual property allocation model in which content is produced but the intellectual property rights are released so that anyone can use the work in electronic form. (*See* green open access and gold open access.)

Open source: Refers to software or documentation that is made available for free use and redistribution provided that certain license provisions are followed.

Pedagogically oriented technical writing: Writing that is focused on teaching. (*See also* professionally oriented technical writing and theoretically oriented technical writing.)

Periodical: Any journal, magazine, or newsletter that is published at some regular rate.

Plagiarism: The representation of others' ideas as your own.

Pre-writing: (*See* Brainstorming.)

Press release: A short statement that is sent to local newspapers and online news services to announce some particularly meaningful event. Also called a news release.

Proceedings: A published transcript of papers presented at a conference.

Professional book: (*See* Trade book.)

Professionally oriented technical writing: Writing that serves the needs of various professionals. (*See also* pedagogically oriented technical writing and theoretically oriented technical writing.)

Pseudocode: A generic name for any code syntax that resembles a programming language but is not intended to be compiled and executed.

Public domain: The state in which the copyright of intellectual property (e.g., writing, graphics, video, sound) has expired.

Publish: A legal term meaning to make the document available to the public, either freely or for a fee.

Referee: One who does refereeing.

Refereeing: The process of reviewing a paper for consideration of publication in a journal or magazine.

Reflowable: In a word processor, text editor, or online publishing system, when displayed content automatically wrap words to the next line as the user changes the window size and thereby relocates the right margin of the page. (*Contrast with* nonreflowable.)

Royalty: A percentage of the net sales revenue of intellectual property.

Scientific writing: Includes experimental research and associated documentation and the scholarly publications that emerge from that work.

Second-person writing: Writing in which the reader is addressed directly. (*See also* first-person writing and third-person writing.)

Social networks: Communities of interests that are enabled by some unifying Internet technology.

Soft DRM: A type of digital rights management for electronic content in which only copying of the file is prohibited. (*See also* hard DRM.)

SWOT analysis: Strengths, weaknesses, opportunities, threats analysis. A study of internal and external factors and positive and negative forces.

Technical reports: Documents that are prepared for supervisors, subordinates, peers, customers, clients, and various government agencies.

Theoretically oriented technical writing: Writing that explicates theoretical and applied research. (*See also* pedagogically oriented technical writing and professionally oriented technical writing.)

Third-person writing: Writing from the perspective of the author as observer. (*See also* second-person writing and third-person writing.)

Trade book: A book that is intended for practitioners. Also called professional book.

Transmittal letter: A letter that accompanies another artifact, such as a résumé, in response to a job advertisement or a defective product return. Also known as a cover letter.

Tuckman Model: A model of team formation that posits that teams can dramatically change from one form to another over time.

Universal quantification: Use of the words "all," "every," "always," and their equivalents. (*See also* existential quantification.)

Vanity press: A commercial publishing firm that underwrites its publications costs by charging the author for the right to publish, rather than investing in the production costs.

White-hat hacker: A computer enthusiast who plays the role of an evil hacker for the purposes of discovering system vulnerabilities and reporting them.

Wiki: A special kind of website with access control mechanisms and version control.

Writing homogenization problem: When a document has clearly been written by multiple authors because of the different writing styles.

Index

Page numbers followed by f & t denote figures and tables, respectively.

Printed in the United States
by Baker & Taylor Publisher Services